MISCELLANEOUS PUBLICATIONS
MUSEUM OF ZOOLOGY, UNIVERSITY OF M̶I̶C̶̶ ̶̶N, NO. 113

Bacula of North American Mammals

BY

WILLIAM HENRY BURT

ILLUSTRATIONS

BY

WILLIAM L. BRUDON

ANN ARBOR
MUSEUM OF ZOOLOGY, UNIVERSITY OF MICHIGAN
MAY 25, 1960

CONTENTS

ILLUSTRATIONS

PLATES

BACULA OF NORTH AMERICAN MAMMALS*

SEVENTEENTH century morphologists were aware of the presence of a baculum (*os priapi*, Ruthenknochen) in certain mammals. Thus, Perrault in 1666 (see Perrault, 1733) was one of the first, if not the first, to mention this bone. Daubenton (1758–1767), according to Retterer and Neuville (1913) was the first to give a comprehensive account of it: "sur les 50 espèces de mammifères dont Daubenton a décrit l'os pénien, il y a 17 types de carnivores." In addition to the carnivores, he listed the bone for primates, bats, moles, pinnipeds, and rodents. Pallas (1767) described and figured the bone for three mustelids and the beaver and later (Pallas, 1778) he described and figured it for the beaver, ground squirrel, and lemming. De Blainville (1839) included accounts of more than 40 species in his Osteographie, and Carus and Otto (1840) described several species. G. Cuvier (1846:208–09) added to the growing literature on the baculum.

The work up to this time was almost entirely on Old World forms. Maximilian (1862) mentioned or gave short descriptions and figures of the bones of eight carnivores and four rodents from North America. Strangely, he stated of the muskrat (p. 171–72) "Penis des Mannchens ohne Knochen."

After a lapse of some 25 years, Retterer (1887) studied the development of the bone in rodents, Arndt (1889) made a study chiefly of the carnivores, and Gilbert (1892) summarized the literature to that date in a paper dealing exclusively with the baculum. In his classical work on the rodents, Tullberg (1899) described the bone for 31 kinds, many of which were figured. About half of these were New World species. Lothar Pohl initiated a series of studies that resulted in publications on the Mustelidae (1909) and carnivores and pinnipeds (1911), among others. The mustelids also received brief attention from Retterer and Neuville (1913) and a more extensive treatment from Pocock (1918). The latter, following the lead of Thomas (1915), also studied the bacula of the Procyonidae (1921) and the Sciuridae (1923). Although not the first to recognize the significance of this element in the classification of mammals, Pocock stressed its importance, particularly in his study of the squirrels (1923). Vinogradov (1925) followed shortly with a study of the Dipodidae and Zapodidae, in which he stressed the importance of this element in classification.

The first really comprehensive study of the bacula of all mammals was that of Chaine (1925). Most of the species treated are of the Old World, but some New World kinds are included. The literature on the subject is assembled in one place and descriptions are given for all kinds known to

* Funds for the publication of this monograph were derived from the income on the endowment of the Horace H. Rackham School of Graduate Studies, and made available by the Executive Board of that School as Project R No. 32—Museum of Zoology.

him at that time. This is a most useful work for anyone studying the bacula of mammals.

Argyropulo (1929) studied the bone, as well as the soft parts, in the Murinae, Tobinaga (1938) published on the bacula of several Asiatic mammals and Ognev in his work on the mammals of the USSR describes and figures the bacula of many kinds. Recently, Didier has had a series of papers (15 in Mammalia) which, combined, constitute the most up-to-date summation of our knowledge of the baculum. Unfortunately, Didier gives few references and little interpretation. In some of the recent large works on mammals, such as Weber and Traité de Zoologie, the baculum is treated briefly.

In this country, no one seems to have considered the baculum worthy of study, probably owing to scarcity of material, until Ruth (1934) published a paper on the development of the bone in the rat and I followed with a study of the bacula of *Perognathus* and *Dipodomys* (Burt, 1936). Since then, a number of short papers, each on one or two species of North American mammals, have appeared in the literature. In addition, studies of genera or families have been made by Wade and Gilbert (1940) on squirrels, Burt and Barkalow (1942) on woodrats, Hamilton on microtines (1946) and vespertilionid bats (1949), White (1953 a and b) on chipmunks, Krutzsch (1954) on Zapodidae, Krutzsch and Vaughan (1955) on additional vespertilionid bats and *Tadarida*, and Dearden (1958) on microtines.

In the present study, I have excluded the bats and the cats. The American insectivores, to my knowledge, have not been investigated for this element. It appears in some of the Old World forms.

I have not treated all groups equally. In those where I believe this character is important in showing relationships, I have given it extra consideration. Furthermore, I have not treated the os clitoridis in detail. It is usually a small bone, often with the appearance of an undeveloped baculum of the same species. Owen (1868:699) mentioned the presence of an "ossicle" in females of the seal and polar bear. Since then, many authors have mentioned this element. It is known to occur in the Chiroptera, Primates, Carnivora, Pinnipedia, and Rodentia. Simokawa (1938) and Layne (1954) have summarized the literature, and Layne has studied the element in the Sciuridae. He found it in all that he examined (8 genera, 28 species). I suspect that the bone is present in most, if not all, of the species that have a baculum.

All measurements are in millimeters. I have thought it unnecessary to indicate this after each measurement. Averages are in parentheses following the range. Many samples are small and ages are unknown, therefore, statistical treatment seems inadvisable. Orientation, when known, is with reference to the erected organ.

MATERIALS AND ACKNOWLEDGMENTS

The collection of bacula now in the Museum of Zoology has been accumulating since 1930 when I first started saving them. Many others have since contributed to this collection. Our staff, especially Emmet T. Hooper and graduate students, have made a special effort to save this part of each specimen. To them I am most grateful. A few of those who have kept us in mind through the years may be mentioned, with the knowledge that I am probably omitting some—to those, my sincere apologies. From Western United States, where this all started, my thanks to Laurencce M. Huey, Luther Little, Arthur G. Barr, Seth B. Benson, Murray L. Johnson, and William L. Jellison. Previous students who have continued to supply material are Burton T. Ostenson, Frederick S. Barkalow, Jr., Harold E. Broadbooks, William O. Pruitt, Edmund A. Hibbard, H. F. Quick, and Andrew Starrett.

The material is preserved in two ways: (1) dry, with the tissues removed and (2) cleared, stained, and preserved in glycerin. It is best to have both kinds of preservation. Dry material is more easily and accurately measured, and the character of the bone is more evident. Cleared and stained material gives the orientation and the relationship of the bone to surrounding tissues.

A substantial amount of the material used for this study was collected on expeditions supported by the Faculty Research Fund (Project No. 466) and by a Grant from the Horace H. Rackham School of Graduate Studies.

SPECIES ACCOUNTS
FAMILY URSIDAE

From the small sample at hand, and from illustrations in the literature, it appears that differences in the bacula of bears are slight. Didier (1950) has given a good account of them. I have but two species represented in the adult form, *Ursus americanus* (Pl. I,g) and *U. sitkensis* (Pl. I,h). Through the middle, the bones are triangular in cross section. At the ends, they are oval. From the enlarged basal end, the bone tapers gradually to near the tip where it again enlarges slightly. The bone may be nearly straight or it may curve slightly upward or downward. Individual variation seems to be primarily in length and thickness of the bone. In three specimens of *americanus* from Michigan, the longest bone is the thinnest. The one specimen of *sitkensis* is heavier and thicker than any of the *americanus*, but is similar in other respects. An os clitoridis is present in *Ursus* and *Thalarctos*.

Measurements of three bones of *americanus* from Michigan and one of *sitkensis* from Baranof Island, Alaska, are, in that order: length, 150.1, 140.4, 166.8, 156.5; height of base, 12.4, 11.1, 10.9, 15.3; width of base, 9.1, 9.1, 7.9, 14.6.

FAMILY PROCYONIDAE

The genera *Procyon, Nasua, Potos, Bassariscus* and *Jentinkia* are represented in our collections. Close relationships are indicated between *Procyon* and *Nasua,* but *Potos, Bassariscus* and *Jentinkia* each has a quite distinctive baculum—an indication of more distant relationships. I do not have a representative of *Bassaricyon,* nor do I find an illustration of its baculum. Hollister (1915:149) made the following statement: "os penis small, 32 mm. in length, slightly bowed, and much less distinctly bilobed anteriorly than in *Procyon* or *Euprocyon.*" This would indicate a terminal end similar to that in the raccoon and, therefore, closer relationships with *Procyon* than shown by *Potos, Bassariscus* or *Jentinkia.* Pocock (1921) described and figured the bacula of the Procyonidae in some detail. He found this character, along with many others, to be of value in indicating relationships. There is considerable diversity in structure of bacula of different genera. This argues for distinct genera possibly by Lower Miocene times, although some are not represented in the fossil record until later.

Procyon lotor.—This raccoon has the largest baculum of any of the procyonids. It has been described and figured many times, and is best known, perhaps, of all the bacula. From the enlarged base, the bone curves gently upward for two-thirds to three-fourths its length, then curves rather abruptly downward. The shaft gradually tapers from base to near the tip where it broadens to end in two condyle-like projections (Pl. II. i). There is some variation in cross-sectional outline of the base, which is usually about equal in height and width, and in the sharpness of the curves. I see no geographic trends in the available material. Age variation is principally in length and thickness of the bone. An os clitoridis is present (Rinker, 1944).

The bone has been used by tailors as a ripping tool as well as an instrument for taking out basting; the proximal end was sharpened to a fine point (Jaeger, 1947).

Measurements of 18 adult specimens from Florida (12), Texas (2), Michigan (2), and North Dakota (2) are: length (taken along a line as nearly parallel to the shaft as possible; some error here), 93.4–111.5 (104.1); height of base, 5.8–10.3 (8.3); width of tip, 5.7–8.7 (7.1).

Procyon cancrivorus.—The baculum of this species (Pl. II, j) resembles that of *lotor,* but the shaft is nearly straight (there is a slight curvature) and more delicate (at least in the one specimen available). Didier's (1950) figures and measurements indicate a thick, heavy bone. The two species of *Procyon* are readily distinguished by this bone alone.

Measurements of one specimen from British Guiana are: length, 106.3; height of base, 6.3; width of tip, 5.7.

Nasua narica.—The illustrations by Pocock (1921), Chaine (1925), and Didier (1950) are all of South American representatives (probably *Nasua nasua*). Pohl's (1911) illustration of the bone of his *"Nasua rufa"* seemes to be *narica,* and the one he calls *"narica"* appears to be the South American species (T.8, figs. 52–54). I have three specimens from Guatemala (2) and Arizona (1). The shaft may curve slightly upward or be nearly straight just anterior to the enlarged base (Pl. II, h). In any event, it tapers gradually to near the tip where it expands into a bilobed terminus which is flat on the dorsal side. On the ventral side, just posterior to the lobes, is a small median projection. In the South American representatives figured, and in our collections, the terminus is similar except that there are no lateral lobes and there is no ventral projection; they are similar to *Bassariscus astutus* in this respect. The female has an os clitoridis (Layne, 1954).

Measurements of two adults from Guatemala (1) and Arizona (1) are: length, 87.6, 87.5; height of base, 5.7, 6.5; width of tip, 7.3, 8.0.

Bassariscus astutus.—This ringtail has the simplest baculum of any of the North American procyonids. It has been figured by Blainville (1839), and Pocock (1921). From the enlarged base, the bone tapers rather evenly to near the distal end where it flattens dorsoventrally with a lateral expansion (Pl. II, f). Viewing the bone from its lateral aspect, it curves gently upward then downward toward the tip. One specimen from Michoacán is 50 mm. long; height of base, 4.8; width of tip, 4.2 The same measurements for a specimen from Arizona are: 43.6, 3.0, 3.1.

Remarks.—Except for size, the bone is similar to those from South American representatives of *Nasua* (see above).

Jentinkia sumichrasti.—The shaft may be nearly straight (Lönnberg, 1911) or curved downward from the enlarged proximal end (UMMZ 109692). Base higher than wide. There is a condyle-like enlargement on the tip; proximal to this, on the ventral side, are two tubercles (Lönnberg, 1911) or short ridges with a median depression (UMMZ). Lönnberg's specimen from Costa Rica was "43 mm." long.

Measurements of one specimen from Oaxaca are: length, 47.0; width of base, 2.7; height of base, 4.3; width of tip, 1.8.

Remarks.—The baculum of *sumichrasti* differs markedly from that of *astutus* and supports the thesis of generic distinction.

Potos flavus.—The bone was described and figured by Pocock (1921). From the slightly enlarged basal end, the shaft tapers gradually to near the tip where it branches into four condyle-like processes. The smaller processes point upward and outward, the larger downward and outward (Pl. II, g). The shaft curves gently upward then downward toward the tip in one specimen, and is nearly straight in another. The four processes on the distal end are unique in the Procyonidae.

Measurements of the one bone from Chiapas are: length, 74.6; dorso-ventral diameter of base, 5.1; width of tip, 5.2. Another, from Guatemala, measures: 79.0, 7.3, 6.0.

FAMILY MUSTELIDAE

The diversity in this large family of carnivores is displayed in the bacula as well as in other morphological characters. Within each genus there is considerable uniformity, but between genera there is little. The genera *Mephitis* and *Spilogale* show similarities, but each of the other genera has its own peculiar type of baculum. This, to me, indicates fairly early separation at the generic level. I have not seen a baculum of *Conepatus*, but Didier's (1948) illustration of one from the South American *C. chilensis* shows it to be quite different from either *Mephitis* or *Spilogale*.

GENERA *Mephitis* AND *Spilogale*

Mephitis mephitis.—The striped skunk has the simplest baculum of any mustelid examined. It is a slender spicule, slightly enlarged at the basal end, which tapers gradually to the distal end. There is usually a gradual curve in the shaft (Pl. III, b), and the distal end is sometimes slightly enlarged and recurved, but not noticeably.

Measurements of nine adult specimens from Michigan (2) and Montana (7) are: length, 20.1–22.8 (21.2); height of base, 0.8–1.3 (0.98); width of base, 0.8–1.3 (1.0). The series is not large enough to demonstrate geographic variation.

Spilogale putorius.—The baculum is a thin, slightly curved spicule. The basal end is sculptured (Pl. III, a). In this respect it somewhat resembles certain species of *Mustela,* but here the resemblance ceases.

Measurements of two adult specimens from Alabama (1) and Florida (1) are, respectively: length, 22.1, 18.8; height of base, 1.4, 1.2; width of base, 1.2, 1.0.

Remarks.—Two young (?) specimens from Montana and Texas have simple, rounded bases, and are considerably shorter than the two listed above. Their lengths are 15.1 and 11.7, respectively.

GENUS *Mustela*

Within this genus the bacula vary in size, rugosity of the basal end, and sharpness of curvature on the distal end. However, they all possess the ventral groove along the distal two-thirds or three-fourths of the shaft and all have slight asymmetry at the distal end. They are recognizable as belonging to *Mustela* and nothing else. In the North American species, *frenata* and *erminea* are similar in the gently sweeping curvature of the distal end, while *rixosa, vison,* and *nigripes* each has a sharply curved, hook-like distal end (Pl. III). There is some geographic variation, as will be pointed out in the species accounts. Age variation seems to be a matter of size and rugosity, especially of the basal end. The female has a small os clitoridis.

Mustela erminea.—This species has the simplest baculum in the genus. From the base, which may be simple, round, and of smaller diameter than the middle of the shaft, as shown in the illustration (Pl. III, c), or rugose and enlarged as shown for *frenata* (Pl. III, d), the shaft curves gently downward then upward to the tip. It is the left side of the tip that is extended to give it asymmetry. Four specimens from Minnesota and three from South Dakota all have small, rounded bases. The specimens from South Dakota may be of young animals as they are considerably shorter (14.0–16.6 mm.) than those from Minnesota (20.9–21.2 mm.). In a series of 32 specimens from British Columbia, ten have small rounded bases (some obviously young) and 22 have rugose bases similar to that shown for *frenata* (Pl. III, d). I believe this is not entirely owing to age as some with rugose bases are shorter than some with smooth bases.

Measurements of four specimens from Minnesota are: length, 20.9–21.2 (21.1); height of base, 0.5–0.9 (0.75); width of base, 0.5–0.9 (0.8). The same measurements for 17 specimens from British Columbia are: 20.1–24.9 (22.2); 0.7–2.2 (1.7); 0.5–1.7 (1.35).

Mustela frenata.—The bone in young *frenata* is similar to that in adult *erminea.* As the bone matures it becomes noticeably heavier and rugose at the basal end, and the distal end curves more abruptly (Pl. III, d). The groove may extend for two-thirds the length of the bone in young specimens, but usually does not reach more than half way down the shaft in adults. However, it has the typical characteristics of the genus.

Wright (1947) weighed the dried bacula and found that in juveniles they ranged from 14 to 29 mg.; in adults, from 53 to 101 mg. There was no overlap between the two groups. The change from a juvenile type to an adult type apparently is quite rapid, probably within a month (Deanesly, 1935). Wright found that all animals in active spermatogenesis possessed the adult-type baculum.

In a series of seven young to adult specimens from North Dakota the distal parts curve less abruptly than do those from Michigan, Virginia, and Alabama. It will require more specimens to demonstrate if this is a real geographic difference.

Measurements of eight adults from California (1) North Dakota (3), Michigan (3), and Alabama (1) are: length, 22.9–27.6 (25.6); height of base, 3.0–4.1 (3.4); width of base, 1.3–3.1 (1.9).

Mustela rixosa.—The baculum of *rixosa* differs primarily from those of *erminea* and *frenata* in having the distal end curved sharply back to form a hook (Pl. III, e). It resembles *vison* and *nigripes* in this respect. It is also the smallest bone found in the genus. Viewed from above, the distal end is somewhat offset to the right. The proximal end is rugose, and the groove may extend for the entire length of the shaft or for only the distal half. This variation does not seem to be correlated with age or geographic distribution. *M. rixosa* may be identified by this bone alone.

Measurements of eight adults from North Dakota (1) and Michigan (7) are: length, 13.6–15.1 (14.2); height of base, 1.5–2.1 (1.8); width of base, 1.0–1.3 (1.1).

Mustela nigripes.—The black-footed ferret possesses a typical bone for the genus. The terminal end hooks sharply backward, in which respect it is similar to that of *rixosa* and *vison*. It is also offset to the right as is that of *rixosa*. The proximal end is a simple, laterally flattened base (Pl. III, f) in one specimen and rugose with a collar similar to that shown for *M. vison* (Pl. III, g) in the other. If this is an age difference, full length of the bone must first be attained, then broadening and rugosity follow. The groove extends more than half way down the shaft. Incidentally, except for being smaller, the bone is similar to one from a laboratory *M. putorius*.

Measurements of the two specimens from Montana and North Dakota are, respectively: length, 36.5, 36.7; height of base, 5.5, 2.2; width of base, 4.1, 1.2.

Mustela vison.—The shaft of the bone in *M. vison* may be nearly straight for its proximal two-thirds or it may curve gradually upward (Pl. III, g). There is considerable variation in this respect. The terminal portion curves sharply back to form a hook. The left lip on the "hook" is always higher than the right. Asymmetry is also found in the basal portion where the ridge or collar which develops in old animals is always more anteriorly situated on the right side than on the left. The distal part of the shaft is not offset as in *nigripes* and *rixosa*. The groove extends varying distances from the tip; usually it is rather deep for half the length of the shaft, then it disappears or shallows as the base is reached. The base is higher than wide in adults. (One specimen of 126 measured has the base wider than high.)

Elder (1951), working with bacula from ranch mink of known age, was able to arrange them in two age classes, juveniles less than one year of age and mature mink over one and one-half years of age. He found weight to be more reliable than length of bone.

Injury to the bone occurs in a small percentage of individuals. In 1668 juveniles and 352 adults (a total of 2020), six show evidence of having been broken and repaired and two are otherwise malformed.

Relationships, as shown by the baculum, are with *M. nigripes* and *M. rixosa,* and more remotely with *frenata* and *erminea,* in that order.

Measurements of 126 adult specimens from North Dakota are: length, 41.2–52.5 (47.98); height of base, 4.2–7.8 (6.27); width of base, 2.9–5.9 (4.63). Specimens from British Columbia (5), Montana (2), and Michigan (1) fall into the size range given for the North Dakota specimens except that one specimen from British Columbia measures 8.0 mm. in height of base. One specimen from North Carolina, which appears to be an adult, is 39.9 mm. long.

GENERA *Gulo, Lutra, Taxidea, Martes,* AND *Eira*

Gulo luscus.—The bone in this species tapers gently from the expanded base to near the tip where there is a shallow groove on what I have interpreted as the ventral side. At the very end, the lateral crests thicken and join in the middle, leaving a very small opening at their bases. Opposite the groove (which is open in a young specimen, remotely similar to that in the mink) is a projection or keel (Pl. III, i). The shaft has a very slight double curvature.

Measurements of two adult specimens from British Columbia are, respectively: length, 83.5, 87.2; height of base, 8.8, 7.8; width of base, 5.3, 5.8.

Lutra canadensis.—Friley (1949a) figured and described the baculum of the river otter in detail. In general, it is a rather heavy bone with an expanded, rugose basal end and a deep urethral groove at the distal end, which is upturned rather sharply and laterally flattened (Pl. III, l). The entire bone, from lateral view, forms what Friley termed a "thin S" in shape. With a series of 83 bacula from spring-caught otters, Friley placed 42 in his "older adults" class (over 93.5 mm. in length), 13 in his "younger adults" class (87.0–93.5 mm.), 25 in his "immature" class (68.0–86.9 mm.) and 3 in his "very young" class (below 68.0 mm.). An os clitoridis is present in the female (Scheffer, 1939).

Lengths of 55 adults from Michigan, as given by Friley are: 87.7–160.4 (94.92 ± 4.46).

Relationships within the family, as indicated by shape and the deep groove, seem to be with the genus *Mustela.*

Taxidea taxus.—The baculum of the badger is unique among the mustelids, as well as the carnivores as a group. The enlarged, slightly higher than wide base tapers into a heavy shaft which has a slight left twist to and including the somewhat flattened, upturned distal end (Pl. III, m). There is no grove.

Measurements of eleven specimens from Idaho (1), Montana (3), Michigan (4), South Dakota (1), Washington (1), and one unknown are: length, 92.2–106.3 (98.7); height of base, 10.4–12.6 (11.4); width of base, 8.0–11.3 (9.4). An os clitoridis is 9.3 mm. long. It is shaped something like a curved spoon. There is no resemblance to the male element (Hoffmeister and Winkelmann, 1958).

Martes americana.—In young specimens, the diameter of the shaft is about the same from the base to near the tip. It curves gently upward, then down, then up again at the tip, which is expanded into two processes that are curved to the left. As the bone matures, the base expands and the tips of the two processes fuse, leaving an opening similar to the eye of a needle (Pl. III, j). Except for the distal end, the bone is fairly symmetrical. An indistinct groove is present near the tip on what I interpret as the ventral side.

Measurements of 48 adults from British Columbia are: length, 35.0–41.9 (38.8); height of base, 2.5–4.3 (3.3); width of base, 2.2–3.7 (2.9).

Martes pennanti.—Except for size, the baculum of the fisher is similar to that of the marten. The distal end is broader and more spatulate with the opening near the middle (Pl. III, k). I do not have sufficiently young material to show the development of the distal end. The genus *Martes,* and the species within the genus, may be recognized by this bone alone. The female has an os clitoridis.

Measurements of three adults are, respectively: length, 102.3, 100.0, 96.9; height of base, 5.8, 7.5, 6.0; width of base, 6.1, 7.2, 7.3.

Eira barbara.—We have one specimen belonging to this species. It is laterally flattened from the base to near the tip, which is dorsoventrally flattened and horseshoe-shaped on the top surface (Pl. III, h). I know of no other baculum like it. The bone is quite different from the one figured by Didier (1947:139).

Measurements of the one bone from Panamá are: length, 78.4; height of base, 6.5; width of base, 3.6; width of tip, 5.6.

FAMILY CANIDAE

The shape of the baculum is fairly uniform in members of the Canidae. This indicates to me a rather close relationship between genera—much closer than that found in the Mustelidae or Procyonidae. M. E. Miller (1952), among others, described and figured the bone for the domestic dog. The shaft either bows upward, downward, has a double bow, or is nearly straight. A deep urethral groove extends from the base (*Canis, Vulpes, Alopex*) or from near the base (*Urocyon*) to near the distal end of the bone. A small keel or sharp crest extends along the dorsal part of the bone opposite the groove. The apex may be round and pointed or slightly enlarged and flattened. There is considerable individual variation in these details, but the general structure of the bone is always indicative of the family.

A few young examples indicate that in the early formation of the structure there are two small bones lying parallel. As development continues, these elements fuse to form a single bone (Jellison, 1945). An os clitoridis is present in the female.

Urocyon cinereoargenteus.—The baculum of the gray fox (Pl. II a) differs from those of other North American canids in that the urethral groove stops short of the basal end, and a well-developed keel is present on the dorsal aspect of all available adult specimens (11). If turned over, the bone resembles a miniature boat.

Lengths of eleven specimens from Lower California (1), Nevada (2), New Mexico (1), Michoacán (1), Florida (1), Georgia (1), Alabama (2), North Carolina (1), and Michigan (1) are 44.0–60.8 (53.1).

Vulpes fulva.—Four specimens from Michigan are fairly uniform in shape and size (Pl. II, b). The urethral groove is relatively broad and extends to the base. The distal end is slightly expanded and broader than high in three; in one it is round. Two have low but distinct keels on the dorsal surfaces; the others have rather sharp ridges. Lengths of the four range from 53.2 to 54.8 (53.9).

Alopex lagopus.—I have one adult (Pl. II, c) and one young specimen. The urethral groove runs the entire length of the bone, although it is shallow at the distal end, which is relatively broader than in other canids treated here. The dorsal ridge is fairly sharp for the basal half, then rounds off distally. The bone curves to the right, but is nearly straight from lateral view. The one specimen from Alaska is 59.3 mm. long.

Canis lupus.—Three specimens from Minnesota (1) and Michigan (2) are available. They all bow upward and all are fairly symmetrical. The deep, broad groove extends about four-fifths the length. The distal end is rounded and slightly expanded on the extreme tip (Pl. II, e). Lengths of the three specimens are, respectively, 83.7, 95.7, and 99.3. They are considerably larger than those of the coyote.

Canis latrans.—In a series of 14 bones from subadult to adult animals, some curve upward, some downward, some have a double curve, and four have lateral curvatures as well. Five have extremely sharp dorsal crests, the others slightly rounded ridges. The urethral groove extends two-thirds to three-fourths the length of the shaft (Pl. II, d). Lengths of eleven specimens from Montana (5), Michigan (2), New Mexico (3) and México (1) are 62.5–88.7 (76.8).

FAMILY OTARIIDAE

I have but one adult bone of the California sea lion (*Zalophus californianus*) to represent this family of pinnipeds (Pl. I, i). The bone is rugose at both ends. Similar to those of the bears, it is triangular in cross section except at the two ends. The extreme distal end is expanded into a roughly circular disc. The bone measures: length, 176; height of base, 15.8; width of base, 24.5; height at extreme tip, 22.3; width of tip, 19.9.

Remarks.—Scheffer (1950) figured the baculum of the fur seal (*Callorhinus ursinus*) for various age groups. He (Scheffer, 1949) also figured the os clitoridis, and Sierts (1950) reported the same for *Z. californianus*.

FAMILY ODOBENIDAE

The walrus (*Odobenus rosmarus*) possesses the largest baculum of any known mammal; it was figured by O. J. Murie (1936). From the base, the bone tapers gradually to near the distal end where it expands to form a small knob. The midsection is roughly triangular in cross section. One specimen, an adult from Hudson Bay, Canada, measures: length, 547; height of base, 46.3; width of base, 50.5; height of tip, 41.0; width of tip, 34.0. Loughrey (1959) gives the following for 14 adults: length (mm.), 308–535 (467.7); weight (gms.), 118.0–768.3 (459.8). Except for size, the bone is not markedly different from that of the sea lion.

FAMILY PHOCIDAE

A baculum from a young male *Erignathus barbatus* shows the characteristic pinniped features of triangular cross section of middle of shaft and a slight expansion of the distal end. The bone is 86 mm. long; the base is 10 mm. high and 14 mm. wide. The adult bone would, I suspect, be considerably larger. Scheffer (1949) figured a small os clitoridis from the harbor seal (*Phoca*).

FAMILY APLODONTIIDIAE

Aplodontia rufa.—The baculum in this species is quite characteristic, and differs markedly from those of all other rodents (Pl. I, d). It is a fairly thin bone, concave ventrally, convex dorsally, and with the basal end smaller than the shaft or tip. The bone gradually widens toward the distal end where it forks; each branch terminates in a condyle-like structure. The base is also notched. It was figured about natural size, possibly slightly enlarged, by Tullberg (1899).

Measurements of one bone from California are: length, 20.8; height of base, 1.3; width of base, 3.0; width across distal branches, 6.7.

Remarks.—No close relationships with other rodents are indicated by the baculum. An os clitoridis is present in the female (Scheffer, 1942:443).

FAMILY SCIURIDAE

The classification of the squirrels has long been a controversial subject. I shall not review the history here because it has been done by a number of authors (Thomas, 1915; Pocock, 1923; A. H. Howell, 1938; Ellerman, 1940; Bryant, 1945; Simpson, 1945; White, 1953 a; Moore, 1959). The most thorough historical reviews are by Ellerman and Moore. Thomas (*ibid.*) first recognized the difficulty of properly evaluating relationships on cranial characters and proposed the use of the baculum as supplementary evidence. Pocock (*ibid.*), following the lead of Thomas, again stressed the value of using this bone. Mossman *et al.* (1932) studied the male reproductive tract in several members of the family and concluded that more fundamental characters were to be found here than in the cranium or external features. A. H. Howell (*ibid.*), concerned primarily with North American squirrels, recognized the structure of the baculum as an aid to classification, but did not use it in his final analysis. Ellerman (*ibid.*) was concerned with the squirrels of the world and based his classification chiefly on cranial characters. Ellerman's philosophy is pretty well summed up in the following quotation (p. 267). "But if this character [baculum] is given such importance, I fail to see how fossil forms are to be considered; and it seems that if cranial and dental characters have been used primarily for classification since the days of Linnaeus one cannot be blamed for wishing to continue to give more importance to these characters than to an external character which has only been definitely verified in a very small percentage of named species and races, and found to be subspecifically variable in at least one case." One might ask where the fields of genetics, physiology, and others of this nature would be today had they abandoned their studies because they could not be applied directly to fossils. Further, I think we have made some progress since the days of Linnaeus. New techniques should not be discarded because they are not universally applicable at the moment. The baculum is not an external character. The fact that a character is variable does not rule against its use. What character in living things is not variable? It is important to know the limits of variability, then to evaluate the character properly as it applies to a population.

Wade and Gilbert (1940) described 17 and figured bacula for 16 North

American Sciuridae. Bryant (1945) gave very brief descriptions of the bacula of 15 species of North American squirrels, and White (1953a, 1953b) figured 21 species, primarily chipmunks. Moore (1959) considered the baculum
along with cranial characters in his recent study. I should like, again, to
stress the importance of this bone, along with other characters, in any determination of relationships within the Sciuridae. I am aware of the possibility
of placing too much emphasis on a single character, and of becoming biased
in judging its value. However, any additional aid that we can discover
should be used, especially in such a complex group as the squirrels.

The diversity in structure of the baculum reaches its height in this
family of rodents. There seems to be no one characteristic that is common
to all genera. This diversity seems a bit strange when one considers the
external similarity throughout the group. The superficial resemblance of
the genera *Sciurus* and *Tamiasciurus* is not reflected in the structure of the
bacula. The entire reproductive tracts are different (Mossman *et al.*, 1932).
The Old World species (*S. vulgaris*), the New World species of *Sciurus*
(except two species, *aberti* and *griseus*), and the genus *Microsciurus* all have
characteristic bones that differ only slightly from species to species (Pls.
VIII, IX) in size and configuration. That of *Tamiasciurus* is no more than
a spicule (Layne, 1952). Again, in the North American genus *Glaucomys*,
the two species, *sabrinus* and *volans*, are difficult to distinguish on cranial
and external characters, yet the bacula bear no similarity (Pl. IV, a, b, c.).
The explanation for this disparity is not apparent in our present state of
ignorance. If one may be permitted to speculate, he can call upon conver-
gent or parallel evolution of the superficial (more adaptive?) structures (if
they evolved) and divergent evolution of the bacula, and possibly other in-
ternal (less adaptive?) structures. If members of the family Sciuridae, as we
know it today, had a common ancestry, some such explanation seems
necessary. I believe this would not violate the tenets of the present-day
evolutionist.

GENUS *Marmota*

The baculum of *Marmota flaviventris* has been figured and described by
Wade and Gilbert (1940) and by J. A. White (1953a). Pocock (1923), ear-
lier, illustrated the bone of *Marmota marmota,* as did Didier (1953:71).
These figures, particularly those of Didier, show the similarity of the Old
World marmots and those of North America in respect to this element. Of
the American members of the Sciuridae (except for *Tamiasciurus*), *Marmota* has the simplest baculum. It has a relatively large basal end which
measures, in greatest diameter, about one-third the length of the bone. From
this expanded base the bone tapers rapidly into the shaft, which curves
generally upwards and ends either in a blunt point or a slightly expanded

tip with a shallow dorsal concavity. The border of the tip may be smooth, rugose, or have from two to five small, blunt, projections, less well developed than those in the genus *Citellus* (Pl. IV, h, i). Of three specimens of *Marmota monax* and five of *M. flaviventris* that I have before me, no two are precisely alike. One specimen of *M. flaviventris* (No. 87344) has a distinct keel on the dorsal one-third of the shaft; the keel ends just short of the smooth tip. There is a slight asymmetry to all the specimens available. Their total lengths range from 3.5 to 4.7 (4.2), a small bone for the size of the animal. A larger series might show significant differences between species, but my series is too small to indicate any such distinctions, particularly in view of the great variability that they display. The hoary marmot (*M. caligata*) is not represented.

The female has a small os clitoridis. Two specimens measure 1.8 each in greatest length.

GENUS *Cynomys*

The bacula in members of this genus are similar to those found in some species of the genus *Citellus*, as was pointed out by Wade and Gilbert (1940). Considering the diversity in the latter genus, as now constituted, *Cynomys* and *Citellus* might well be combined if one were relying on the characters of the bacula alone (Pls. IV, V, VI). Actually, bacula of *Cynomys* appear to be closer to those of some members of the subgenus *Citellus* than are those of other subgenera such as *Otospermophilus* (see pp. 20–22). A. H. Howell (1938, Pl. 13, M) gave a rather poor illustration of the bone in *Cynomys*. He also described it briefly (p. 38).

Within the genus *Cynomys*, the two subgenera (*Cynomys* and *Leucocrossuromys*) are distinguishable in the limited material available. Four specimens of *Cynomys* have five or six teeth each on the terminal end (Pl. IV, g), if one includes a small projection back on the shaft which is not present in *Leucocrossuromys* (Pl. IV, e). Each of the two specimens of the latter subgenus has eight projections. In addition, the terminal denticulate portion is broader and more flared than in *Cynomys*. One specimen of *Cynomys* (No. 96071) has low rounded projections similar to those in *Marmota*. A median projection on the ventral side of the terminal end is similar to that found in most, but not all, of the members of *Citellus*. This structure is not present in *Marmota*.

Measurements of four specimens of *Cynomys ludovicianus* are: length, 4.4–5.1 (4.65); width of basal end, 1.2–1.5 (1.35); width of distal end, 1.3–1.5 (1.4). Two specimens, one each of *Cynomys leucurus* and C. *gunnisoni*, measure 4.2 and 4.4 in length; 1.7 and 1.9 in width of basal end; 1.7 and 1.5 in width of distal end.

An os clitoridis is present in both groups. One of C. *ludovicianus* is 2.5

mm. and one of *C. leucurus* is 2.2 mm. long. They have the appearance of undeveloped bacula.

GENUS *Citellus*

The bacula in members of the genus *Citellus* characteristically have tooth-like projections around the borders of the expanded distal ends, which are roughly spoon-shaped. The projections, mostly sharply pointed, are directed dorsad. They may form a continuous row or there may be a gap in the median area. On the ventral side of the distal end of the bone is either a median knob-like projection or a keel in all examined except *C. annulatus*. The enlarged basal end tapers rapidly toward the center of the shaft which again broadens and curves upward distally. The shaft usually appears to have a slight twist in it. These same characteristics are found in the genus *Cynomys*, and the bacula of *Marmota* are not very different. There is no ventral projection or keel in the latter, and the tooth-like projections are but slightly indicated. The marmots, prairie dogs, and ground squirrels, thus, constitute a fairly unified group. No other North American squirrel possesses this type of baculum.

Pocock (1923) placed these in the subfamily Marmotinae, chiefly on the basis of the baculum. This is consistent with the classification by A. H. Howell (1938) who, although recognizing that the baculum might furnish characters of taxonomic worth, used primarily characters of the skull in his final analysis. I suspect that he had too few specimens and that he did not study them critically. Simpson (1945:79) listed the above, plus the genera *Tamias* and *Eutamias*, in his Tribe Marmotini, which he indicated is equivalent to Pocock's subfamily. From characters of the bacula, the indication is that *Tamias* and *Eutamias* do not belong in this group.

I cannot agree with White (1953a:560) that "The baculum in *Tamias*, in general plan of structure, resembles that in *Spermophilus* [*Citellus*] and *Cynomys* of the tribe Marmotini of Simpson." A generally spoon-shaped distal end is the only thing common to the two.

In addition to three American species, *annulatus, mexicanus,* and *tridecemlineatus,* Pocock (*ibid.*) figured the baculum of the Old World species *C. mongolicus*. Didier (1952) recently illustrated *C. fulvus* and *C. citellus*; Wade and Gilbert (1940), *lateralis, variegatus, tridecemlineatus, franklini, richardsoni,* and *spilosoma*; while White (1953a) figured *C. armatus*. All have denticulate margins of the expanded distal discs. Howell's (1938) figures are mostly too small to show details of structure. Layne (1954) illustrated the os clitoridis for ten species.

Citellus variegatus and *beecheyi.—*These squirrels are currently given subgeneric rank (*Otospermophilus*). The bones from members of the two species are indistinguishable, but they differ distinctly from those of all other members of the genus (Pl. V, a, b). *C. atricapillus* is not represented. The basal ends of bacula of members of *variegatus* and *beecheyi* are relatively less enlarged than are those of members of other species of *Citellus*. From the proximal end of the bone, there is a gradual tapering of the shaft until it again expands, as it curves upward, into the distal scoop which is concave on the dorsal surface, and normally has two projecting teeth on each side. There is a median projection ventrally. The distal end also is enlarged relatively less than in representatives of other species. The narrowly expanded distal end and the reduced number of teeth are similar to conditions found in *Cynomys*, particularly in *ludovicianus* (Pl. IV). I would, therefore, place the subgenus *Otospermophilus* next to *Cynomys* in relationships as shown by the bacula. A. H. Howell (1938:43; pl. 13, G, H, I) thought it close to *beldingi*, subgenus *Citellus*.

Seven specimens (6 *variegatus* and 1 *beecheyi*) range in length from 4.0 to 4.5 (4.1); width of base, 0.6–1.1 (0.8); and width of distal end, 0.6–1.1 (0.8). An os clitoridis is 2.0 mm. long.

Citellus richardsoni.—Citellus richardsoni and *C. elegans* are currently treated as subspecies because A. H. Howell (1938:75) indicated intergradation between the two in Gallatin County, Montana. If they do intergrade we have a rather unusual situation in regard to the structural differences in the bacula—differences greater than normally found between species. Three specimens of *C. r. richardsoni*, one each from Alberta, Montana, and North Dakota, have either very small projections or none at all on the expanded distal ends (Pl. VI, c, d.). This, in itself, is unusual for the genus. Contrariwise, three specimens of *C. r. elegans*, one each from Montana, Wyoming, and Colorado, have from 8 to 11 distinct tooth-like projections on the distal expansion, as is characteristic of the genus (Pl. VI, a, b). The row of teeth in the *elegans* series is continuous, and there is a terminal ventral keel similar to that in *columbianus*. The chief differences, other than size, between the *elegans* series and those of *columbianus* are in shape of the terminal disc and number of tooth-like projections (fewer in *elegans*). Both are currently placed in the subgenus *Citellus*.

Even though my sample is small, the material is from widely scattered areas and probably indicates the type of baculum found in each population. If true, there must be a fundamental difference in the two gene pools. There is no perceivable difference in the measurements of the two series. They are grouped here as follows: length, 2.9–3.1 (3.0); width of base, 0.8–0.9 (0.85); width of distal end, 0.9–1.3 (1.15). Numbers of tooth-like projections in the *elegans* series, 8–11. The female has an os clitoridis. Three specimens measure 1.6–2.0 in length.

Citellus beldingi.—"The baculum from a specimen of *Citellus beldingi oregonus* is 3.5 mm in length; its shaft is broad at the base, slightly curved, and narrowed toward the tip, which is shaped like a spoon with a crenulate border. The apex of the shaft appears as a short process projecting from the lower surface of the terminal disk." (Howell, A. H., 1938:41; pl. 13, Q) From the small figure, and the above description, it would appear that in general configuration the bones in members of this species most closely resemble those in members of *columbianus*.

Citellus columbianus.—The enlarged basal ends of the bacula in representatives of this species are nearly round in cross section. In a representative bone, the shaft narrows rather abruptly, with a slight twist, then expands again into the distal scoop-like end which is studded with tooth-like projections, and is more nearly horseshoe-shaped than in any other in the genus. The teeth number from 12 to 20; in all but six specimens they are unequal in number on the two sides. There is a median, keel-like projection on the ventral border of the distal end (Pl. V, 1). The baculum is distinct, in a subtle way, from those of all other species, particularly in the shape of the distal end.

Measurements of 18 specimens from Montana are: length, 4.3–5.5 (5.2); width of base, 0.7–1.3 (1.1); width of distal end, 1.4–1.8 (1.6). The female has a small os clitoridis; two specimens measure 2.2 mm. each in length.

Citellus undulatus.—One specimen from Alaska, all that is available for this species, is about equal in size to a baculum of *variegatus,* but it differs in that it has a more widely expanded distal end with more tooth-like projections (seven). The continuous row of projections occupies only the front half of the disc. The proximal shoulders of the disc, which merge into the shaft, have slightly crenulated edges. There is a ventral, median knob on the distal end. The proximal end is expanded and nearly circular in cross section.

Measurements of the one specimen are: length, 4.0; width of basal end, 1.1; width of distal end, 1.2.

Citellus tridecemlineatus.—The bacula in members of this species have characteristically broad bases which taper rather abruptly into the narrow shafts. The shafts, relatively longer than those of most other species except *mexicanus,* expand distally and curve dorsally into broad scoops which are bordered by several sharp tooth-like projections. There is a ventral median projection on each (Pl. VI). Howell (1938:42) states that ". . . there is no process" in his specimen. Normally, in bacula of members of this species, each has small sharp, tooth-like projections, variable in number, exterior to the main row. There may be none, as in two specimens before me, one on each side, or occasionally as many as seven when a double row of teeth gives the bone the appearance of a rooster's rose comb. This was the kind that

Pocock (1923:235) figured. The number of teeth varies from 10 to 24 in the series before me. Another variable in this series, which appears to be geographic in nature, is that each specimen from Michigan and one from North Dakota, as well as the one figured by Pocock (*ibid.*), from Minnesota, has a median gap (Pl. VI, 1), whereas those from Kansas, Montana, and South Dakota have continuous rows of teeth (Pl. VI, j), as do those from Nebraska (Wade and Gilbert, 1940). A. H. Howell (1938, pl. 13, N) figured one with a gap in the middle, but gave no locality.

Measurements of 14 specimens from Michigan (7), South Dakota (2), North Dakota (1), Montana (2), Kansas (1), and Oklahoma (1) are: length, 4.0–5.2 (4.5); width of base, 13 specimens, 1.0–1.5 (1.2); width of distal end, 12 specimens, 1.5–1.8 (1.6). An os clitoridis is present in the female.

Citellus mexicanus.—I have but two bacula, one from Texas and one from Tamaulipas, México. They are indistinguishable from the eastern representatives of *C. tridecemlineatus* (Pl. VI, e ,f). Each has 13 tooth-like projections on the distal end, and each has a median gap. Bacula of this species were also figured by A. H. Howell (1938, pl. 13, D), and by Pocock (1923: 235).

Measurements of two specimens are: greatest length, 4.1, 4.8; width of base, 0.8, 1.4; width of distal end, 1.5, 1.6.

Citellus spilosoma.—The bacula are similar to those of members of *C. tridecemlineatus* from the western part of the range (Pl. VI, g, h). However, they have shorter, heavier shafts and wider distal ends. The number of tooth-like projections varies from 14 to 18 in four specimens.

Measurements of four specimens from Arizona (3) and Texas (1) are: length, 3.6–4.0 (3.8); width of base, 1.3–1.6 (1.4); width of distal end, 2.0–2.2 (2.1).

Remarks.—The three species, *tridecemlineatus, mexicanus,* and *spilosoma* were placed in the subgenus *Ictidomys* by A. H. Howell (1938). Curiously, members of *tridecemlineatus* from the eastern part of the range are nearest *mexicanus,* and those from the western part of the range approach *spilosoma* in characters of the bacula. The close relationship of the three species, based primarily on other characters, is also indicated by the bacula.

Citellus tereticaudus.—The bacula of members of this species are not unusual in any particular. The enlarged basal end of the bone tapers rapidly and asymmetrically into the shaft; this almost immediately expands in an asymmetrical way into the terminal spoon-like portion which is studded with a continuous row of 8 to 10 tooth-like projections. A small, rounded knob projects forward medially and ventrally (Pl. V, g, h). I see no very close resemblance to the bone of *variegatus,* as did A. H. Howell (1938:45;

pl. 13, R). Although his figures indicate a resemblance, specimens before me do not bear this out.

Measurements of three specimens from Nevada are: length, 2.6, 2.7, 2.8; width of base, 1.0, 0.9, 1.0; width of distal end, 1.1, 0.9, 1.0. An os clitoridis is 1.4 mm. long.

Citellus franklini.—The basal ends of the bacula of representatives of this species are oval in cross section, wider than high. The taper from the base to the shaft is rather abrupt and asymmetrical, mostly along the right side. The shaft continues rather uniform in thickness for the central one-third of the bone, then abruptly expands into the broad, disc-shaped distal end. Between the shaft and the tooth-like projections, on each side, is a rounded lobe-like protuberance. Similar structures are indicated in *spilosoma* and *columbianus*, but they are most prominent in *franklini*. The tooth-like projections, numbering 12 to 15, are in a continuous row (Pl. V, i, j). A median projection, sometimes rounded, sometimes slightly bifid on the end, extends beyond the main bone ventrally and distally. In this respect, and in the prominent lobes, the bone differs from those of all other species examined. Howell (1938:43) stated that "the terminal disk . . . [is] . . . without pronounced crenulations on its margin." I suspect that the tooth-like projections were broken off in the cleaning process; this likely was true with some of his other specimens also. His figure (pl. 13,P) is too small to show this feature.

Measurements of three specimens from Indiana (1), North Dakota (1) and South Dakota (1) are: length, 4.5, 4.5, 4.8; width of base, 1.5, 1.9, 2.1; width of distal end, 2 specimens, 1.9, 2.2. One os clitoridis is 2.2 mm. long.

Citellus annulatus.—The basal ends of the bacula in members of this species are distinctly oval in cross section; they are wider than high. The bone tapers rapidly into the slightly curved shaft which expands abruptly into the broad distal end. The distal spoon-like portion is bordered by a continuous row of tooth-like projections (24 in one and 25 in the other of two specimens from Colima and Jalisco, México). The pointed projections curve inward at the sides, but around the front of the disc they curve outward, except for one projection near the middle that curves inward. This situation is not seen in any other representatives of *Citellus* examined. Further, the anterior disc is broadest of any here considered. There is no ventral projection on the distal end (Pl. V, e, f).

Pocock (1923:235) figured the bone for "*C. leursi*" (a *nomen nudum*) from Jalisco, México. This most certainly is *C. annulatus.*

Measurements of the two specimens are, respectively: length, 5.2, 5.2; width of base, 2.0, 2.1; width of distal end, 2.8, 2.7.

Remarks.—The forward-curving tooth-like projections and the smooth ventral surface of the distal end, without a median projection, are found in no other bacula of *Citellus* examined.

Citellus lateralis.—Bacula in representatives of this species are small for the size of the animals, and fairly simple. The bone has a typical broad base, nearly circular in cross section, that tapers asymmetrically into the shaft which flares at the distal end into a nearly symmetrical spoon-like portion studded with 6 to 10 sharp, evenly spaced, tooth-like projections (Pl. IV, k). A blunt knob projects from the ventral surface of the distal end (Pl. IV, j), which is less prominently curved than those of most other species of *Citellus*. An os clitoridis is about 1.7 mm. long. Howell's figure (1938, pl. 13, J) of this bone is too small to show anything of its character.

Measurements of six adults from Arizona (1), Colorado (2), Utah (1), Idaho (1), and Montana (1) are: length, 2.5–3.0 (2.7); width of base, 0.7–0.8 (0.75); width of distal end, 0.6–1.1 (0.9).

GENUS *Ammospermophilus*

Ammospermophilus leucurus.—Members of this genus have small, delicate bacula. The broad base of the bone tapers, almost entirely on the right side, abruptly into the slender shaft. Distally, the shaft broadens, gradually on the right side and rather suddenly on the left, into a circular ladle- or dipper-like portion with the sides (or wings) curved back toward the base. On the dorsal edge of the circular, distal expansion are 13 to 17 sharp, tooth-like projections which curve outward. There is no ventral knob or keel on the distal end (Pl. V, c, d). The bone is quite distinct from those of the genus *Citellus*. That of A. *harrisi* (Howell, 1938, pl. 13, K, L) is indistinguishable from the above. Variation in the limited material available seems to be primarily in the number of tooth-like projections and the extent that the wings curve back toward the base. One os clitoridis is about 2.0 mm. in length.

Measurements of two specimens from Nevada are, respectively: length, 2.0, 2.1; width of base, 1.0, 1.0; width of distal end, 1.6, 1.7.

Remarks.—A. H. Howell (1938:44) thought that the baculum in *harrisi* resembled those of the genus *Citellus*. I see no very close relationship other than the tooth-like projections. The bone is quite distinct; its characteristics may be added to those of skin and skull in support of full generic rank for these squirrels.

GENUS *Tamias*

In this monotypic, strictly North American genus, (*T. striatus*) the baculum is unique in many respects (Pl. VII, k, l). The shaft is round, or nearly so, at the base. As it tapers distally, it is somewhat compressed on the sides. The distal end is best described as spoon-shaped, upturned to form an obtuse angle with the shaft, concave dorsally and convex ventrally, where there is a keel. The keel is prominent in some, slightly developed in others. Also variable is the acuteness of the point on the distal end and the extent of asymmetry developed as the left edge of the "spoon" tapers into the shaft proper.

The relationship with other sciurids, as Wade and Gilbert (1940) and White (1953a, fig. 7) have pointed out, is probably closest to *Citellus*. The baculum of *Tamias* has no teeth on the edge of the distal part, and some representatives of *Citellus* approach this condition (*C. richardsoni*). Certainly, there is no evidence from the baculum that *Tamias* and *Eutamias* should be placed in the same genus. The female has an os clitoridis.

Measurements of 31 specimens from Michigan (27), Virginia (3), and Maine (1) are: length, 3.5–4.7 (4.15); width of base, 0.4–0.7 (0.58); width of tip, 0.6–0.8 (0.7).

GENUS *Eutamias*

North American members of the genus *Eutamias,* as now constituted, possess bacula which are uniform in general structure, and which differ from those of all other squirrels (Pl. VII). They vary in proportions of parts and in detail of structure from species to species, but basically they constitute a unified group. The shaft may be round or dorso-ventrally flattened at the base; it may be nearly straight or with one or two curves; the distal part (tip) may have a strong or weak central keel on its dorsal aspect; and the angle that the tip forms with the shaft varies. Then, there are size and age differences.

White (1953b) described 16 and figured bacula of 15 species of *Eutamias.* Wade and Gilbert (1940) previously described and figured the bone for *amoenus* and *minimus.* Inasmuch as these were studied rather critically by White (*ibid.*), I shall discuss them briefly here. All that have been examined have an os clitoridis.

Eutamias alpinus.—"Shaft thin; keel low, . . . angle formed by tip and shaft 135°: . . . base slightly wider than shaft; shaft short, 2.17 mm." (White, 1953b:616, fig. 1)

Eutamias minimus.—Shaft slender, slightly curved; base slightly enlarged, nearly circular; tip forms an obtuse angle with shaft (Pl. VII, g). The specimens are quite uniform in character. In my material, they are closest to those of *townsendi* (not figured) and *dorsalis* (Pl. VII, i, j).

Measurements of 25 specimens from Michigan (7), Montana (8), South Dakota (3), Nebraska (4), and Wyoming (3) are: length, 3.2–4.1 (3.96);

length of tip, 0.7–1.1 (0.9); width of tip, 0.2–0.5 (0.37); height of base, 0.2–0.5 (0.34); width of base, 0.4–0.5 (0.37).

Remarks.—Three specimens from South Dakota average 4.0 mm. in length; seven from Michigan average 3.5; eight from Montana average 3.4 and seven from Nebraska and Wyoming average 3.6. A larger series might show a geographic difference in size of baculum.

Eutamias townsendi.—Similar to that of *minimus*, but slightly shorter and thicker in region of main shaft; base somewhat flattened, wider than high; angle of tip and shaft slightly less obtuse than in *minimus*. The bone was figured by White (1953b, fig. 3).

Measurements of one young specimen from Washington are: length, 3.2; length of tip, 0.9; width of tip, 0.3; height of base, 0.3; width of base, 0.5.

Eutamias dorsalis.—General shape of bone similar to that of *minimus*, but larger over all; base very slightly wider than high (Pl. VII, i, j).

Measurements of nine specimens from New Mexico (5), Arizona (1), and Nevada (3) are: length, 3.1–4.6 (4.0); length of tip, 0.8–1.2 (1.0); width of tip, 0.2–0.4 (0.34); height of base, 0.3–0.5 (0.4); width of base, 0.4–0.6 (0.53).

Eutamias sonomae.—According to White (1953b, fig. 4), the bone is similar to those of *minimus* and *townsendi* except that it has a lower keel on the tip and the ridges on either side of the tip are more pronounced. The obtuse angle that the tip forms with the shaft is the same as in *townsendi*.

Euamias amoenus.—Baculum relatively stockier than that of *minimus*; base wider than high, may or may not be bifid on dorsal surface; keel low and not prominently notched (Pl. VII, h). White (1953b, fig. 5) figured what is apparently a young individual. The tip seems to develop faster than the base.

Measurements of ten specimens from Montana (6), Idaho (2), Washington (1), and Oregon (1) are: length, 2.9–4.6 (3.5); length of tip, 0.7–1.6 (1.1); width of tip, 0.3–0.5 (0.4); height of base, 0.1–0.5 (0.34); width of base, 0.4–0.8 (0.58).

Eutamias merriami.—"Shaft thin, keel low, . . . angle formed by tip and shaft . . ." (White, 1953b), same as in *townsendi, sonomae,* and *amoenus;* shaft long. White (*ibid.*:621) gives the length of shaft as 4.88 mm.

Eutamias cinereicollis.—Judging from the description and figure given by White (1953b:624, fig. 13), this species is second to *quadrimaculatus* in length of baculum. Although the measurement he gives for the length of shaft (4.88 mm.) is the same as that given for *merriami* and less than that given for *quadrimaculatus* (4.35 to 5.28), the illustration shows the bone to be considerably larger than either of the others. The shaft has a slight S-curve, and the tip forms an obtuse angle with it (145 degrees). The keel on the tip is relatively low and broadly notched. Size alone, it seems to me, is less important than general configuration. I, therefore, suggest that *cinereicollis* might better fit with the *alpinus, minimus, townsendi, sonomae, amoenus, dorsalis, merriami* group than with *quadrivittatus* and *ruficaudus*.

Eutamias quadrimaculatus.—From the measurements given by White (1953b:624) this seems to be the longest of the bacula represented in his material (4.35 to 5.28 mm.). The shaft is relatively straight, and the angle it forms with the tip is less obtuse than that of *cinereicollis*. I would place it between *cinereicollis* and *ruficaudus* in the series.

Eutamias ruficaudus.—Baculum relatively heavy throughout; shaft with S-curve; base nearly twice as wide as high, usually bifid; tip and distal part of shaft form angle slightly greater than 90 degrees; keel on tip prominent and notched at base (Pl. VII, f).

Measurements of two adults from Montana (1) and Washington (1) are, respectively: length, 5.4, 4.6; length of tip, 1.6, 1.6; width of tip, 0.6, 0.6; height of base, 0.5, 0.5; width of base, 0.8, 1.0.

Eutamias quadrivittatus.—Shaft relatively heavy with definite S-curve; base distinctly wider than high, usually bifid; keel on tip with distinct notch (Pl. VII, e).

Measurements of 12 specimens from Colorado (1), New Mexico (10), and Nevada (1) are: length, 3.4–4.6 (4.35); length of tip, 1.1–1.5; (1.35); width of tip, 0.4–0.5 (0.48); height of base, 0.4–0.6 (0.5); width of base, 0.7–0.9 (0.8).

Eutamias bulleri.—The bone in this species is the longest of any represented in my material; the nearest to it in size is that of *ruficaudus*. The base is distinctly wider than high and is deeply notched (Pl. VII, a, b). The keel on the tip is relatively low and straight-edged; other than this, the general configuration is similar to that of *quadrivittatus*. The tip forms an obtuse angle with the slightly curved shaft.

Measurements of five specimens from Durango are: length, 5.1–5.7 (5.3); length of tip, 2.0–2.2 (2.06); width of tip, 0.5–0.8 (0.66); height of base, 0.8–1.1 (0.9); width of base, 1.2–1.5 (1.36).

Eutamias umbrinus.—The baculum in this species has a relatively short thick shaft, the base of which is more than twice as wide as high. There is a distinct bend near the middle. The tip makes an angle of about 90 degrees with the distal part of the shaft (Pl. VII, d). Relationships, as indicated by the bacula, are with *panamintinus, speciosus,* and, according to White (*op. cit.*), with *palmeri*.

Measurements of two specimens from Utah are: length, 3.1, 3.3; length of tip, 1.0, 1.2; width of tip, 0.3, 0.5; height of base, 0.3, 0.4; width of base, 0.7, 0.9.

Eutamias panamintinus.—From White's description and figure (1953*b*, fig. 16), the bone has a short, sharply curved shaft with "base markedly widened." It appears to be intermediate in characters between those of *umbrinus* and *speciosus*. White (*op. cit.*) gives the length of shaft as 2.17 mm., while that for *speciosus* is given as "2.11 to 3.17 mm." His illustration indicates that the bone for *panamintinus* is distinctly longer than that of *speciosus*.

Eutamias speciosus.—The bone in this species is the most distinctive in the genus. It has a short thick shaft with a distinct bend, and a flattened, bifid base that is more than twice as wide as high. The tip is approximately at right angles to the shaft (Pl. VII, c). The configuration is most closely approached by bacula of *panamintinus* (White, 1953*b*, fig. 16) and *umbrinus* (Pl. VII, d).

Measurements of three specimens from California are: length, 2,6, 2.7, 2.8; length of tip, 1.3, 1.3, 1.2; width of tip, 0.4, 0.5, 0.5; height of base, 0.4, 0.4, 0.3; width of base, 0.9, 0.9, 0.9.

Remarks.—The supposed close relationship of *speciosus* and *quadrivittatus* is not borne out by the bacula. Howell (1929) considered this a subspecies of *quadrivittatus,* but Johnson (1943) accorded it specific rank.

GENUS *Sciurus*

There are two distinct types of bacula in members of the genus *Sciurus* as now constituted (see Wade and Gilbert, 1940). One type, found in all but two species (*griseus* and *aberti*), is recognizable at a glance, and is distinct from all other bacula known (Pls. VIII and IX). The basal portion of the shaft is circular or nearly so in cross section. The shaft tapers distally, with an apparent twist, to its smallest diameter which I am calling the neck. At this point the shaft usually curves dorsally and expands into a broad circular disc which is concave on the right side and convex on the left. Ventral to this expanded disc is a definite spur and, in some, a supplementary small spur in front. Variation in this type of baculum is primarily in size, proportions, depth of concavity, presence or absence of a supplementary spur, presence or absence of a small tuberosity proximal to the spur, and the outline of the posterodorsal edge of the expanded disc, pointed or rounded. A low dorsal keel on the shaft is, I believe, an age character; it is present in some, absent in others.

Species possessing this general type of baculum have been variously subdivided into subgenera by different authors. Inasmuch as no two agree in the arrangement, it seems quite clear that they have been arbitrary in many instances. I believe that too much weight has been given the character "presence or absence of a small premolar," and this has led to confusion in determining real relationships. I am not concerned here with subgenera, but only with a linear arrangement of the species represented by bacula to indicate possible relationships as shown by that element. I am tentatively placing those with this type baculum in three subgroups as follows: (1) *Sciurus niger, S. oculatus, S. arizonensis,* and *S. carolinensis.* (2) *Sciurus poliopus, S. colliaei, S. nelsoni, S. socialis, S. variegatoides,* and *S. yucatanensis.* (3) *Sciurus deppei, S. negligens, S. granatensis, S. (Microsciurus) alfari,* and *S. alleni.*

The second type of baculum found in the genus is represented by those of *griseus* and *aberti.* The bacula of these two species are not only larger than the others, but the expanded distal end forms a vertical, elongated, flat keel which is pointed both distally and proximally. There is a small ventral spur on that of *aberti,* a slight tuberosity at the position where one would expect the spur in *griseus.*

In the various species considered here, length ranges from 8.8 to 16.9 mm. An os clitoridis is probably present in all females. It is small and undeveloped, but recognizable as belonging to *Sciurus.*

The illustrations (Pls. VIII and IX) are turned over, dorsal side down and ventral side up, that the more characteristic side (right) of the distal expansion might be shown. I trust that this will not lead to confusion.

Sciurus niger.—The baculum in this species is relatively large; the shaft tapers but slightly, and the neck just back of the distal expansion is fairly thick. The spur is rather blunt, but prominent; there is no supplementary spur, nor is there a tuberosity back of the spur. The expanded distal part, in most specimens, terminates posterodorsally in a rather sharp point which forms an acute angle with the shaft (Pl. IX, a). Also, the edges of the terminal disc may be flared out or directed inward nearly to a horizontal position.

Measurements of 15 specimens from Michigan (8), Iowa (1), Nebraska (1), Alabama (1), and Georgia (4) are: length, 11.4–15.4 (12.6); length of expanded tip, 2.5–3.5 (2.9); height of tip, 2.9–3.6 (3.25); height of base, 2.2–3.3 (2.9); width of base, 1.9–2.8 (2.4). One female element is 3.7 mm. long.

Sciurus oculatus.—In this species, the shaft tapers to a narrow neck, just back of the expanded distal end, and the spur is blunt instead of coming to a point; this is particularly true of the one illustrated (Pl. VIII, c). The other specimen available is not quite as extreme, but shows the same general configuration. Some specimens of *niger* approach this condition, and those of *alleni* are suggestive of it. The posterior part of the disc ends in a definite point and is shaped like a hook. There is a prominent dorsal keel on the shaft.

Measurements of two specimens from Querétaro are, respectively: length, 10.8, 11.5; length of expanded tip, 2.7, 3.0; height of tip, 3.2, 3.2; height of base, 2.5, 2.8; width of base, 1.9, 2.1.

Sciurus arizonensis.—The baculum has no supplementary spur, nor does it have a tuberosity back of the main spur. There is some variation in the way the spur projects from the shaft. The bone figured (Pl. VIII, f) has no notch posterior to its base, and the spur appears to be a continuation of the shaft. Other specimens have a notch there and the spur forms an angle with the shaft. Actually, the bone is quite similar to that of *carolinensis.*

The posterior edge of the disc is either rounded or forms a small point. In the former situation, the angle formed with the shaft approximates 90 degrees; in the latter, there is the appearance of a small hook.

Measurements of eight specimens from Arizona are: length, 11.2–12.2 (11.9); length of expanded tip, 2.4–3.0 (2.7); height of tip, 2.6–3.1 (2.9); height of base, 2.5–2.9 (2.7); width of base, 1.9–2.5 (2.3).

Sciurus carolinensis.—Shaft of bone tapers little for proximal two-thirds; neck, posterior to expanded distal end, relatively thick; distal end forms lesser angle with main shaft than in other species; no supplementary spur or tuberosity posterior to spur, which is recurved and sharp or fairly straight and blunt (Pl. IX, f).

The posterior edge of the disc varies from a rounded, blunt point to a definite hook with a sharp point. A short dorsal keel is present on the shaft just posterior to the neck.

Measurements of 17 specimens from Maine (2), Pennsylvania (1), Michigan (4), Minnesota (3), Iowa (2), Alabama (4), and Mississippi (1) are: length, 9.2–12.3 (10.6); length of expanded tip, 1.8–2.6 (2.3); height of tip, 2.5–3.1 (2.8); height of base, 1.8–2.8 (2.4); width of base, 1.5–2.5 (2.0). One female element is 3.5 mm. long.

Sciurus poliopus.—The bone in this species is rather typical of the group (Pl. VIII, g). The shaft tapers gradually from a broad base, which is slightly higher than wide, to the neck just posterior to the expanded distal end. The spur on the ventral side is prominent, and a small supplementary spur anterior to the main one is present in eleven of the twenty specimens available. Also, a small tuberosity, posterior to the spur, is evident in ten specimens. The posterior edge of the expanded disc is normally rounded off, and the angle formed with the shaft is approximately 90 degrees. A keel may or may not be present on the dorsal part of the shaft.

Measurements of 19 specimens from Michoacán (10), Jalisco (8), and Oaxaca (1) are: length, 10.1–13.8 (11.8); length of expanded tip, 2.6–3.2

(2.9); height of expanded tip, 2.8–3.5 (3.2); height of base, 2.3–3.5 (3.0); width of base, 1.8–3.0 (2.46).

Sciurus colliaei.—The baculum resembles that in *S. poliopus* (Pl. VIII, e), but has a relatively heavier shaft in the two specimens available. The slight differences might come within the range of normal variation. There is a small spur anterior to the main one in one specimen. A tuberosity, posterior to the spur, is present in both. The posterior edge of the disc is rounded; it forms approximately a right angle with the shaft. There is a dorsal keel on the shaft of one specimen.

Measurements of two specimens from Jalisco are, respectively: length, 10.9, 12.1; length of expanded tip, 3.0, 3.2; height of expanded tip, 3.3, 3.5; height of base, 3.3, 3.1; width of base, 2.8, 2.9.

Sciurus nelsoni.—The one representative of this species has a relatively small expanded distal end. It has a large main spur with a distinct supplementary spur anterior to it and a small tuberosity to its posterior. The distal disc is rather deeply concave; the anterior edge is rolled inward, and the posterior edge forms a right-angle point which in turn forms a slightly greater than right angle with the shaft. The illustration (Pl. IX b) shows a tendinous attachment along the ventral surface. This is a common occurrence in all individuals. In old age, this occcasionally ossifies (Wade and Gilbert, 1940).

Measurements of one specimen from Distrito Federal, México are: length, 12.0; length of expanded tip, 3.2; height of tip, 3.2; height of base, 3.0; width of base, 2.5.

Sciurus socialis.—One bone is represented. It is fairly simple, does not have a supplementary spur, but has an indication of a tuberosity proximal to the spur. The expanded distal end is rounded on its posterior edge and forms approximately a right angle with the shaft (Pl. IX, d).

Measurements of one specimen from Oaxaca are: length, 12.7; length of expanded tip, 2.7; height of tip, 3.0; height of base, 3.5; width of base, 2.8.

Sciurus variegatoides.—The baculum in this species (Pl. VIII, h) has a rather large base, tapers to a rather narrow neck, and has a relatively small expanded tip, with the edges curled over in three specimens. There is no supplementary spur, but a definite tuberosity is present just posterior to the main spur. The posterior edge of the disc is rounded and a right angle is formed with the shaft. There is a dorsal keel on the shaft of one specimen.

Measurements of four from Costa Rica and one from Panamá are: length, 11.5–12.4 (12.1); length of expanded tip, 2.4–2.7 (2.6); height of tip, 2.7–3.6 (3.0); height of base, 2.7–3.3 (3.0); width of base, 2.1–2.9 (2.4).

Sciurus yucatanensis.—The one specimen available (Pl. VIII, d) has a relatively high base and small expanded distal end. There is a trace of a supplementary spur, but no indication of a tuberosity posterior to the main spur. The expanded part is deeply cupped on the right side, the dorsal edge is bent over, and is in a transverse plane relative to the vertical axis of the bone. It has a small point on its posterior edge.

Measurements of one specimen from Quintana Roo are: length, 10.6; length of expanded tip, 2.1; height of tip, 2.6; height of base, 3.0; width of base, 2.4.

Sciurus deppei.—The bacula in this species are relatively short and stocky; there is a distinct supplementary spur anterior to the main one, similar to *negligens* in this respect. The posterior edge of the disc forms a sharp point; there is a definite notch between the point and the shaft.

Measurements of two specimens from Costa Rica and one from Guatemala are: length, 9.8, 10.0, 10.0; length of expanded tip, 2.7, 3.0, 2.5; height of tip, 3.0, 3.1, 3.2; height of base, 3.3, 3.6, 2.9; width of base, 2.5, 2.5, 2.3.

Sciurus negligens.—Baculum similar to that of *S. deppei*, but somewhat longer and slimmer (Pl. IX, e). There is a small supplementary spur anterior to the main one on the ventral side. The expanded tip comes to a definite point posterodorsally, and a notch is formed with the shaft. There is a definite dorsal keel on the shaft.

Measurements of one specimen from Tamaulipas are: length, 11.1; length of expanded tip, 2.6; height of tip, 2.8; height of base, 2.5; width of base, 2.3.

Sciurus granatensis.—The baculum is short with a relatively thick shaft and a large expanded distal end. It has a small supplementary spur anterior to the main one. This does not show in the illustration (Pl. VIII, b) because of the angle at which the bone is drawn. The posterodorsal part of the expanded tip is sharply pointed and a notch is formed with the shaft, which has a dorsal keel. The bone would indicate definite relationships with *S. deppei* and *S. negligens*.

Measurements of two specimens from Costa Rica and Panamá are, respectively: length, 8.8, 10.2; length of expanded tip, 3.0, 3.1; height of tip, 3.0, 3.5; height of base, 2.6, 3.0; width of base, 2.2, 2.3. One female element, from Panamá, is 3.7 mm. long.

Sciurus (Microsciurus) alfari.—Baculum small, but not relatively so; about the size and configuration of that in *granatensis* from Costa Rica; typical of those in most of the species of *Sciurus*, except *aberti* and *griseus*; shaft tapers gradually anteriorly to the expanded distal part which is relatively large (Pl. VIII, a), and which terminates posteriorly in a sharp point. The notch formed with the shaft gives the appearance of a hook.

Measurements of one specimen from Costa Rica are: length, 8.8; length of expanded tip, 2.7; height of tip, 3.0; height of base, 2.9; width of base, 2.3.

Remarks.—From this element alone, the relationship with *S. granatensis* is clearly indicated. Certainly, there is no reason to place *alfari* in a genus separate from *Scuirus*. One could make a case for *aberti* and *griseus,* but not for *alfari.*

When J. A. Allen (1895) proposed *Microsciurus* as a subgenus, the characters he gave were general proportional ones, especially an expanded "malar." Goldmen (1912) elevated *Microsciurus* to full generic rank, and added to Allen's brief characterization "simpler dentition than *Sciurus* . . ." In my estimation, both of the above authors placed too much importance on minor characters. I doubt if subgeneric rank is justified. I, therefore, propose that *alfari* be placed in the subgenus *Guerlinguetus* Gray.

Since the above was written, Moore's (1959) classification has been published. He admits the close relationship between *Microsciurus* and *Guerlinguetus* (p. 178), but follows by placing *Microsciurus* in a different subtribe, Microsciurina. I cannot agree with this decision. The only character that he gives as evidence for separation of the two is the extent to which the squamosal extends up the side of the cranium.

Sciurus alleni.—The baculum in this species is average in size and proportions. There is no supplementary spur. One specimen has a small tuberosity just posterior to the prominent spur. The posterodorsal edge of the expanded tip ends in a blunt point (Pl. IX, c) and an acute notch is formed with the shaft, which has a pronounced dorsal keel.

Measurements of two specimens from Nuevo León are, respectively: length, 11.0, 11.0; length of expanded tip, 3.1, 2.9; height of tip, 3.2, 3.0; height of base, 2.8, 2.9; width of base, 2.2, 2.3.

Sciurus griseus.—The bone in this species is the largest (with the possible exception of *S. aberti*) of those represented. The shaft is fairly round and gently curved on the basal half. The distal end has a high flat keel on its dorsal edge; the keel is pointed both distally and proximally. The ventral surface bears a slight tuberosity in the position of the spur on *S. aberti* (Pl. IX, h).

Measurements of two specimens from Los Angeles County, California, are, respectively: length, 15.2, 16.9; length of keel, 5.8, 6.3; height of distal end, 2.2, 2.4; height of base, 2.0, 2.5; width of base, 1.6, 2.3. One os clitoridis is 5.4 mm. long.

Remarks.—Relationships, as indicated by the baculum, are with *S. aberti,* and distantly with other members of the genus. This was pointed out by Wade and Gilbert (1940).

Sciurus aberti.—The rounded shaft of the baculum in this species has a gradual taper distally and a gentle curve ventrally (Pl. IX, g). A flattened, high keel forms the tip of the bone; there is a small spur ventrally, the spur seems to be homologous with that of other members of the genus and serves to indicate intermediacy between *S. griseus* and other species of *Sciurus*. The bone is longer than any other except that of *griseus*.

Measurements of two specimens from Durango and New Mexico are, respectively: length, 15.9, 16.3; length of keel, 4.0, 4.2; height of distal end, 2.9, 2.9; height of base, 2.4, 2.6; width of base, 1.7, 2.3.

Remarks.—Size and configuration of the bone indicate closest relationship with *S. griseus*. The small spur and more rounded keel than that of *griseus* distantly link *alberti* with other species of *Sciurus*.

GENUS *Glaucomys*

The two species in this genus differ so markedly in the character of the baculum that one would likely place them in separate genera if he had only this element on which to base his classification. A comparison of the baculum of *G. sabrinus* with the illustrations and descriptions given by Pocock (1923:242–44, fig. 28 c, d, e) for *Hylopetes alboniger* lead me to the belief that *sabrinus* is more closely related to *Hylopetes* than it is to *volans*. If true, this might indicate a later interchange between Asia and North America for *sabrinus* than for *volans*. The southern distribution for *volans* and the more northerly one for *sabrinus* also supports this theory.

Glaucomys sabrinus.—The bone in this species is relatively short, broad, and doubly curved from lateral view (Pl. IV, a, b). The base is concave from an end view. From the proximal end the shaft narrows slightly then expands into a broad trough-like section that narrows as it curves abruptly upward with a twist of about 90 degrees. The concave side becomes convex and the convex side concave. There is a prominent tooth-like projection at the bend. The distal portion is superficially similar to that of *Sciurus*, and probably indicates distant relationships with that genus. White (1953a) figured and briefly described the baculum of *sabrinus*.

Measurements of 17 specimens from Michigan (9), North Carolina (1), Washington (1), Montana (1), and South Dakota (5) are: length, 6.3–7.3 (6.8); greatest width of shaft, 1.7–2.1 (1.8). A female element, 0.8 mm. long, is a simple nodule with little character.

Glaucomys volans.—The bone in this species is a long, slender, twisted shaft with the basal end but slightly expanded (Pl. IV, c). The shaft is widest at about mid-point where a groove starts and continues to the distal end. The bone terminates in a double condyle-like structure with the groove of the shaft continuing to near the anterior edge of one of the "condyles."

There is no resemblance to other sciurids. Pocock (1923) thought that the bone of *volans* was similar in many respects to that of *Hylopetes*. I cannot agree with him on this.

Measurements of seven specimens from Michigan (6) and Louisiana (1) are: length, 12.1–12.8 (12.6); greatest width of shaft, 1.0–1.1 (1.04). An os clitoridis is 1.0 mm. long.

FAMILY CASTORIDAE

Castor canadensis.—Friley (1949*b*) studied 50 bacula of beaver from Michigan. Of these, he classed 34 as yearlings, 12 in the two- and three-year class, and four as four years or older. Inasmuch as he figured the baculum and described it in detail, suffice it to state that the bone is a fairly simple one with an enlarged base which tapers rapidly into the shaft that ends dis- ally in a blunt, rounded point (Pl. I, f). The shaft has a slight downward curvature, and the base is flattened ventrally. Pallas (1778) gave an excellent illustration for the european beaver.

Measurements of the three classes are from Friley (1949 *b*). Yearling class (34 specimens): length, 22.2–31.5 (26.8); height of base, 3.3–6.35 (4.8). Two- and three-year class (12 specimens): length, 28.0–34.2 (31.4); height of base, 5.6–7.4 (6.6). Four-year-olds and older (4 specimens): length, 32.9–36.6 (34.4); height of base, 7.4–8.6 (7.9).

FAMILY GEOMYIDAE

Members of this family have fairly simple rod-like bacula, somewhat bulbous on the proximal end and pointed on the distal end. The rounded shaft curves gently upward in most of them, but is nearly straight in some. Definite relationships with the family Heteromyidae are indicated. Size of the bone is not necessarily correlated with size of the animal. The largest bone is that of *Thomomys talpoides* and the smallest is that of the large *T. bulbivorus*. In some, the species may be identified from this bone alone (*talpoides*); in others this is not true. The baculum of *Geomys* is so similar to the bone in some species of *Thomomys* that the genera cannot always be separated without recourse to other characters. Geographic variation, pri- marily in size, is apparent in wide-ranging species such as *T. bottae*. Age variation can usually be detected by the size of the proximal bulbous por- tion, as well as by length of bone.

Thomomys bottae.—The bone in this species is fairly uniform, except for size, throughout the range (Pl. X, m, n). It is simple, with a gentle upward curvature of the shaft. The proximal end is bulbous, usually slightly greater in dorsoventral than lateral diameter, and the distal end tapers to a blunt point.

Although my series from any one locality is too small to be significant except in a general way, those from New Mexico (16 specimens of four subspecies, *collis, morulus, paguatae,* and *planorum*) appear to be largest (length, 11.1–15.7; average, 13.4) and those from Lower California (two specimens of *litoris*) smallest (length, 9.7, 10.9 mm). Some of the small high-mountain forms are not represented. From California, 26 specimens of the following races are represented: *bottae* (3), *cabezonae* (3), *infrapallidus* (7), *melanotis* (4), *mohavensis* (3), *neglectus* (2), *pallescens* (1), *perpallidus* (1), and *perpes* (2). One subspecies, *centralis,* with three specimens, is from Nevada. There are 38 specimens of six subspecies from Arizona: *collinus* (7), *comobabiensis* (7), *hueyi* (7), *modicus* (9), *phasma* (1), and *pusillus* (7); and three spcimens representing two subspecies *anitae* (1) and *litoris* (2), from Lower California.

Measurements of the entire series of 86 specimens are: length, 9.1–15.7 (11.9); height of base, 1.0–2.5 (1.7); width of base, 0.9–2.1 (1.4).

Thomomys umbrinus.—Baculum similar in general to that of *bottae,* but smaller and with the base somewhat flattened laterally (Pl. X, k, l). Four specimens of *T. u. burti* and four of *T. u. proximus,* from the Santa Rita Mountains, Arizona, constitute the sample available. The bones of *proximus* are in the size range of *bottae modicus* from the valley to the west of the Santa Ritas. There is a possibility that *proximus* is a *bottae,* and not *umbrinus.*

Measurements of four specimens of *burti* are: length, 9.0–9.7 (9.4); height of base, 1.4–2.2 (1.6); width of base, 1.0–1.7 (1.25). Four specimens of *proximus* measure: length, 10.9–12.0 (11.2); height of base 1.2–1.7 (1.4); width of base, 1.1–1.3 (1.2).

Thomomys bulbivorus.—This large pocket gopher has the smallest baculum of any geomyids examined (if the specimen before me is adult). It is a simple bone with a bulbous basal end, slightly wider than high, that tapers rapidly into the slender shaft (Pl. X, i, j). The shaft curves slightly upward and terminates in a small knob.

The one specimen from Oregon measures: length, 8.5; height of base, 1.5; width of base, 1.8.

Thomomys talpoides.—This species is represented by the largest bacula of any of the geomyids examined. The bulbous proximal end is slightly if any greater in diameter than in some specimens of *bottae,* but the enlarged part extends farther up the shaft. The entire bone tapers gradually from proximal to distal end, and the shaft is slightly curved upward (Pl. X, o, p). I am not certain that all bones measured are of comparable age; those that obviously are young are not included. However, those measured indicate geographic differences in size as well as proportions. I suspect that if sufficient series were at hand from enough localities within the range of the species some definite clines would be apparent (Table 1). Certainly, *talpoides* should be placed at one end of the linear series of *Thomomys*, not in the middle as in Miller and Kellogg (1955).

TABLE 1

MEASUREMENTS OF BACULA OF *Thomomys talpoides* (MM.)

Number UMMZ	Length	Height of Base	Width of Base	Locality	Subspecies
87412	15.2	1.4	1.3	Montana	*trivialis*
87414	16.4	1.6	1.5	"	"
87422	15.1	1.8	1.9	"	"
87424	16.2	1.4	1.1	"	"
87443	17.4	1.5	1.5	"	*bullatus*
87449	20.7	1.9	1.8	"	"
87450	19.3	1.5	1.8	"	"
87454	20.6	1.4	1.6	"	"
87457	19.2	1.4	1.5	"	"
87458	18.8	1.7	1.5	"	"
87463	17.3	2.1	1.8	Wyoming	*bridgeri*
87468	16.7	2.0	2.3	"	"
90259	18.3	1.4	1.5	South Dakota	*nebulosus*
82114	22.3	1.9	2.0	New Mexico	*taylori*
82117	22.6	2.4	2.5	" "	"
82121	23.1	1.9	1.8	" "	"
82130	21.3	2.3	2.1	" "	"
82134	23.8	2.0	1.9	" "	"
82136	22.5	2.0	2.0	" "	"
82137	25.7	2.3	2.1	" "	"
105653	30.2	1.4	1.5	Washington	*tacomensis*

Geomys bursarius.—The bacula in *Geomys*, although similar to some *Thomomys* and some *Cratogeomys*, usually may be distinguished by the dorsoventrally flattened and laterally expanded tip (Pl. X, a, b). Kennerly (1958) described and figured them for *G. bursarius* and *G. personatus.*

Measurements of five specimens from New Mexico (1), Oklahoma (2), and Minnesota (2) are: length, 10.1–11.9 (11.1); height of base, 1.5–2.0 (1.7); width of base, 1.4–1.8 (1.6).

Cratogeomys varius.—The bone in this species has a broad, dorsoventrally flattened base that tapers gradually into a relatively heavy shaft. The upturned shaft terminates in a laterally expanded, flattened tip. The bone is broader than high except for that part just preceding the distal end, which is round in cross section (Pl. X, e, f).

Measurements of seven specimens from Michoacán are: length, 10.0–12.5 (11.2); height of base, 0.9–1.5 (1.3); width of base, 1.8–2.5 (2.0).

Cratogeomys merriami.—Two specimens from México differ slightly in detail. Both are dorsoventrally flattened from the base to near the upturned tip, which is not flattened as in *C. varius*. One specimen has wing-like lateral projections 3.5 mm. back of the tip, the other does not have these. The first specimen also has a slight keel along its dorsal surface. If the few specimens available are representative, the two species of *Cratogeomys* may be distinguished by this bone alone (Pl. X, g. h).

Measurements of the two specimens are, respectively: length, 13.9, 13.8; height of base, 1.4, 1.5; width of base, 2.6, 3.0.

Zygogeomys trichopus.—I have one specimen from Michoacán (Pl. X, c, d). It is similar to that of *Cratogeomys* in that the shaft is flattened and slightly curved. But the base is only slightly wider than high and the tip is not noticeably expanded. The middle part of the shaft is broad and nearly parallel-sided, then it narrows abruptly about 3 mm. from the distal end.

Measurements of the one specimen are: length 12.8; height of base, 1.8; width of base, 2.1.

FAMILY HETEROMYIDAE

With the exception of *Perognathus hispidus,* bacula in members of this family fall into a general pattern. They are simple rods, usually with expanded basal ends, and with tapering shafts that vary in dimensions and amounts of curvature. Relationship with the Family Geomyidae is clearly indicated.

GENUS *Perognathus*

I have previously described and figured the bacula for several species of *Perognathus* (Burt, 1936). The additional material that has accumulated in the interim substantiates my former conclusions concerning relationships within this genus. Three species (*formosus, baileyi,* and *hispidus*) caused Osgood (1909) trouble when he revised the genus. They did not fit nicely into the two subgenera, so he arbitrarily placed *formosus* in the subgenus *Perognathus* and *baileyi* and *hispidus* in the subgenus *Chaetodipus* with the following qualifying statement. "One of these (*formosus*) is a *Perognathus* with strong inclination toward *Chaetodipus;* another (*baileyi*) presents the reverse case; and the third (*hispidus*) must be classed as a *Chaetodipus,* though it is aberrant in some ways." A study of the bacula bears out Osgood's conclusions based on skin and skull characters.

The silky pocket mice of the subgenus *Perognathus* (except *formosus*) have a short baculum with a relatively large, bulbous basal end that tapers rapidly into the slender shaft which turns up at a nearly right angle distally, and terminates in a point (Pl. XI, c–e).

The baculum in members of the subgenus *Chaetodipus* (except *baileyi* and *hispidus*) is relatively longer and more slender than that just described for the subgenus *Perognathus*. The basal portion is slightly bulbous and the distal end is upturned. As the bone is viewed from the lateral aspect, it is roughly sigmoid in outline (Pl. XI, h, i).

In the species *formosus* and *baileyi,* the bone is extremely slender, only slightly enlarged at the basal end, and the shaft has a gentle curve upward. The point does not bend abruptly as in the others described above (Pl. XI, f, g).

The baculum in *hispidus* differs from all others in the genus, as well as in the family Heteromyidae, in having a three-lobed distal end instead of terminating in a point (Pl. XI, j, k).

There is little variation in the bacula of individuals of an age group in a given species. Between related species, variation is principally in size, curvature of the bone, and relative size of the basal end. The size range in the entire genus varies from 4.1 (*longimembris*) to 18.1 (*hispidus*) in length.

Remarks.—My linear arrangement differs from that of Miller and Kellogg (1955) only in that *formosus* and *baileyi* follow the *spinatus* group, and *hispidus* is placed at the end. To be consistent, two new subgenera should be erected to take care of these three species that do not fit. However, this should be based on a totality of characters, not just the baculum. All measurements of length were made parallel to the long axis of the bone, thus, the turned up tip is not included. They are to the nearest tenth of a millimeter.

Perognathus fasciatus.—The one specimen available (Pl. XI, d), from Montana, is typical of the silky pocket mice. The basal end is less bulbous than in *merriami,* or *parvus,* but the general configuration is the same—a gently curving shaft that tapers gradually to the upturned distal end. It measures 7.5 mm. in length and 1.0 mm. in dorsoventral diameter of the base.

Perognathus flavescens.—Two specimens, one young, from Nebraska and Oklahoma are available. The bulbous base is wider than high and the tip turns up to form an obtuse angle with the shaft, similar to that of *fasciatus*. The adult specimen is 7.0 mm. long and the base is 1.0 mm. in height.

Perognathus merriami.—Three specimens from Texas represent this species. The basal end is large; from the base, the shaft arches slightly upward, then downward and up again at the tip (Pl. XI, c).

Measurements are: length, 5.2, 5.8, 6.1; height of base, 0.8, 1.0, 0.9.

Perognathus flavus.—Eight specimens are available from Arizona (2), New Mexico (3), Nebraska (1), and Querétaro (2). They are fairly uniform throughout this wide range. The large bulbous basal end is wider than high; the shaft curves up in the middle and again at the distal end.

Measurements of six specimens are: length, 6.3–7.2 (6.8); height of base, 1.0–1.2 (1.1).

Perognathus longimembris.—I have ten specimens from California and three from Nevada. The bulbous proximal end varies from being wider than high to higher than wide. The bone in this species is the smallest of those represented. General shape is similar to those described above.

Measurements of 13 specimens are: length, 4.1–6.6 (5.2); height of base, 0.4–0.8 (0.5).

Perognathus amplus.—I had one specimen from Arizona when I did my previous study (Burt, 1936). It is typical of the bones of this group; the base is wider than high and the shaft tapers gradually from the bulbous base to the pointed tip. It was 7.1 mm. long and the base was 0.7 mm. in height.

Perognathus inornatus.—Thirteen specimens, nine of which are adults, are available. Similar to *amplus* and others in the group; shaft tapers from enlarged base to pointed upturned tip.

Measurements of seven specimens are: length, 6.0–6.6 (6.4); height of base, 0.6–1.2 (0.8).

Perognathus parvus.—I had none of these previously, but now have 26 from Washington and one from Wyoming. They are typical of the silky-haired pocket mice, and are the largest in this group. The base is nearly circular in cross section (Pl. XI, e).

Measurements of nine adults are: length, 7.5–8.4 (7.95); height of base, 0.7–1.0 (0.87).

Perognathus penicillatus.—Twenty-eight specimens from California (7), Arizona (8), Sonora, including Turner's and Tiburon islands (10), and Texas (3) are available. The base is nearly circular in cross section; the shaft forms a long sigmoid in lateral view and tapers gradually from the slightly enlarged base to the pointed upturned tip; similar to *goldmani* (Pl. XI, h).

Measurements of 27 specimens are: length, 9.0–12.9 (11.0); height of base, 0.5–1.3 (0.8).

Remarks.—The size range is partially a result of age differences, partially, perhaps, geographic. Arizona specimens appear to be largest with a range in length of 11.5 to 12.9 mm. Specimens from Tiburon Island range from 9.5 to 11.3 in length. There are too few specimens, and none are of known age, to attribute variation in size to any one factor.

Perognathus arenarius.—There are 14 specimens from Lower California. Similar in every respect to the bones of *penicillatus* except that the base is slightly higher than wide.

Measurements of seven adults are: length, 9.7–12.1 (10.5); height of base, 0.8–1.0 (0.9).

Perognathus pernix.—We have one cleaned specimen from Sonora and six in glycerin from Sinaloa. The base is higher than wide and the distal end curves upward sharply. It differs from *goldmani* (Pl. XI, i) chiefly in being smaller. Length of the one specimen is 11.1; height of base is 1.0.

Perognathus intermedius.—Three specimens each from Arizona and New Mexico represent this species. The base is nearly equal in height and width, similar to the bone figured for *goldmani* (Pl. XI, i).

Measurements of five specimens are: length, 10.5–12.3 (11.2); height of base, 0.8–1.0 (0.85).

Perognathus nelsoni.—I have two specimens from Texas and one (broken) from Tamaulipas. They are typical of the spiny pocket mouse group. The two from Texas measure, respectively: length, 12.0, 12.3; height of base, 0.8, 0.9.

Perognathus goldmani.—Six specimens are available from Sonora. These are fairly large, sigmoid-shaped bones with a slightly enlarged basal end and an upturned distal end which forms approximately a right angle with the shaft (Pl. XI, i). The base is slightly higher than wide. These were described under the name *artus* previously (Burt, 1936).

Measurements of five specimens are: length, 13.2–14.1 (13.5); height of base, 0.9–1.1 (0.96).

Perognathus californicus.—Most of the 22 specimens available from California appear to be young. The basal part is nearly in line with the shaft which curves gradually upward to near the end where the tip has an abrupt upward turn. It is easily identifiable as a member of the spiny pocket mouse group.

Measurements of four adults are: length, 11.1–11.6 (11.4); height of base, 0.9–1.0 (0.92).

Perognathus spinatus.—This species is represented by the largest series

(58) of any in the genus, all from Lower California, including many of the islands in the gulf of California. The slightly enlarged basal end is nearly circular in cross section and the entire bone is sigmoid shaped in lateral outline.

Measurements of 51 specimens are: length, 9.3–13.0 (10.9); height of base, 0.4–0.9 (0.68).

Remarks.—Here, as in *penicillatus*, I cannot determine whether variation in size is a result of age differences, individual variability, or of geographic position. The largest series from one locality is from Angel de la Guarda Island. Sixteen specimens range in length from 9.6 to 12.0 mm.

Perognathus formosus.—The baculum in this species is a simple, nearly straight spicule with the base slightly enlarged (Pl. XI, f). The tip may have a perceptible upturn in some.

Measurements of three specimens from Nevada are: length, 8.5, 8.7, 9.1; height of base, 0.4, 0.3, 0.3.

Remarks.—Relationships, based on the baculum, seems to be with *baileyi*. Certainly, they are not with any other species of *Perognathus*.

Perognathus baileyi.—The bone in this species is relatively small for the size of the animal. It is a simple spicule similar to that in *formosus* (Pl. XI, g). The shaft is nearly straight and the basal end is slightly enlarged.

Measurements of 15 specimens from Arizona (1), Sonora (1), and Lower California (13) are: length, 8.5–11.0 (9.2); height of base, 0.5–0.8 (6.47).

Perognathus hispidus.—The baculum in this species is the largest in the genus. The somewhat expanded basal end may be higher than wide, the reverse, or nearly circular in cross section. The shaft is nearly straight and terminates in three lobes, two ventral and one dorsal (Pl. XI, j, k). The bone is unique in this respect.

Measurements of eleven adults and sub-adults from Nebraska (1), Kansas (1), Oklahoma (3), and Texas (6) are: length, 14.8–18.1 (16.4); height of base, 1.3–1.9 (1.57).

Remarks.—The peculiar structure on the distal end of the bone sets this apart from all others. Even generic relationships are not indicated by the baculum. This structure is well developed in one half-grown specimen. Identification of the species may be positive with this element alone.

GENUS *Liomys*

In this genus, as in *Perognathus* and *Dipodomys*, the baculum has a bulbous proximal end, a slightly upturned shaft, and a rather abruptly upturned or laterally flattened distal end. However, the upturned distal portion is of less extent than in *Dipodomys* and many of the *Perognathus*.

Differences between species are primarily in the shape of the distal end. *L. pictus* has a definite ventral keel on the extreme tip; *L. crispus* has the distal end slightly flattened dorsoventrally and broadened laterally; *L. irroratus* has a simple tip that is round in cross section. The three species may be recognized by this element alone. Relationships appear to be closer to *Perognathus* than to *Dipodomys*.

Liomys pictus.—The basal, bulbous end is usually higher than wide and tapers gradually into the slightly upcurved shaft. Subterminally, just before the tip turns upward, the shaft is flattened dorsoventrally; immediately following this it is flattened laterally to form a ventral keel on the upturned tip. The species is easily identifiable by this characteristic alone (Pl. XI, b).

Measurements of 26 specimens from Nayarit (3), Jalisco (2), Michoacán (2), Guerrero (4), Oaxaca (9), Chiapas (5), and Vera Cruz (1) are: length, 8.0–10.0 (8.8); height of base, 0.9–1.7 (1.3); width of base, 0.8–1.5 (1.1).

Liomys crispus.—The series of bacula available average slightly smaller than in *pictus*. The base is higher than wide and tapers gradually into the slightly curved shaft. The extreme tip widens perceptibly then comes to an acute point owing to a dorsoventral flattening of the upturned part. The species may be recognized by this characteristic.

Measurements of 13 specimens from Chiapas are: length, 6.9–8.5 (7.8); height of base, 1.1–1.5 (1.3); width of base, 0.9–1.3 (1.1).

Liomys irroratus.—The baculum differs from those of *pictus* and *crispus* in that the basal part is usually nearly spherical: it tapers rather abruptly into the shaft. Also, the distal end is a simple, round, upturned point (Pl. XI, a).

Measurements of 27 specimens from Jalisco (1), Michoacán (2), Guerero (2), Oaxaca (8), Tamaulipas (1), Querétaro (4), Puebla (5), and Distrito Federal (4) are: length, 7.3–8.7 (7.9); height of base, 1.1–1.7 (1.4); width of base, 1.0–1.6 (1.24).

GENUS *Microdipodops*

Microdipodops pallidus.—I have two adults and three young of this species. The baculum is definitely intermediate between those of the silky pocket mice and the kangaroo rats. The large basal part tapers into a shaft that curves moderately upward, but the end is not upturned as in *Dipodomys* and most *Perognathus* (Pl. XII, b). Simpson (1945) placed them in the Dipodomyinae.

Measurements of two specimens from Nevada are: length, 6.3, 6.8; height of base, 1.0, 1.0; width of base, 0.8, 0.9.

All of the species treated here, except *phillipsi*, were described and figured in a previous paper (Burt, 1936). Close relationship with the genus *Perognathus* is indicated by the structure of the baculum. The bone in *Dipodomys* is characteristic of the genus, although some, particularly in the *merriami* group, approach those of certain *Perognathus*. In my previous study I had but one specimen of *spectabilis;* that was a young one. I was not aware of this at the time, but subsequent material indicates that *spectabilis* has the largest bone in the group, and it should be aligned with those that have greatly enlarged basal ends. I have seen no specimens of *D. elator*. Blair (1954) had one specimen and stated that it "is very similar to the baculum of *merriami* and it is quite unlike that of *spectabilis*." In the linear arrangement that follows, I am deviating from that given in Miller and Kellogg (1955) by beginning with *spectabilis* and *phillipsi*, following with the *heermanni* and *ordi* groups, and concluding with the *merriami* group.

Dipodomys spectabilis.—The baculum in this species is the largest in the genus (Pl. XII, j). The bulbous basal end is sculptured in old animals, and is higher than wide. From this end, the shaft tapers gradually to the pointed tip, which is upturned at approximately a right angle to the nearly straight shaft. One specimen (UMMZ 85414) shows injury or disease on the distal end, which terminates in a small knob with a swollen ring just short of the tip. In adults, the species may be identified by this bone alone. Eight specimens (two young) are available.

Measurements of five adults from Arizona (4) and Texas (1) are: length, 15.3–16.1 (15.6); height of base, 1.9–2.8 (2.2); width of base, 1.4–2.0 (1.7).

Dipodomys phillipsi.—The one specimen available differs from all others in the sharp angle that the upturned tip makes with the shaft (Pl. XII, i). In the others, the tip curves upward, but in *phillipsi* the tip angles abruptly from the shaft. The bulbous proximal end tapers gradually into the fairly straight shaft. In general shape, it seems to be nearest that of *spectabilis,* but there is a great size difference. It certainly does not align with *merriami* (Setzer, 1949).

Measurements of the one specimen from Oaxaca are: length, 10.5; height of base, 1.3; width of base, 1.2.

Dipodomys heermanni.—In this species the large, bulbous basal end of the bone is higher than wide. It tapers abruptly into the shaft which continues to taper gradually to the upturned tip (Pl. XII, f). One specimen (L. M. Huey, 10040) apparently had the tip injured and instead of terminating in a gracefully curved point, it is club-shaped. The angle formed by the tip with the main axis of the shaft varies from a right angle to a slightly obtuse one. Comparing the bone of this species with that of *agilis*, Boulware (1943) stated that "The length and curvature of the os penis, or baculum, readily differentiates the two species. The baculum of *heermanni* is longer and curves downward more abruptly at the distal end than that of *agilis*. In adult *heermanni* . . . the bacula are 11.1 mm., 11.6 mm., and 12.0 mm. long, respectively. In adult *agilis* . . . the bacula are 9.8 mm. and 9.9 mm. long."

Measurements of three adults from California are: length, 11.7, 11.8, 12.0; height of base, 2.3, 1.8, 2.0; width of base, 1.9, 1.5, 1.5.

Dipodomys panamintinus.—In my previous paper (Burt, 1936) these were treated under the names *D. mohavensis* and *D. leucogenys*, now considered races of the species *panamintinus*. The baculum is indistinguishable from that of *D. heermanni*. The series before me averages a trifle smaller than the three specimens of *heermanni*.

Measurements of nine specimens from California are: length, 10.0–11.2 (10.7); height of base, 1.3–1.8 (1.6); width of base, 1.1–1.7 (1.3).

Dipodomys ingens.—Similar in configuration to the bones of *heermanni* and *panamintinus*, but possibly slightly larger (Pl. XII, h).

Measurements of three specimens from California are: length, 10.8, 12.4, 13.1; height of base, 1.8, 2.0, 2.1; width of base, 1.6, 1.9, 1.9.

Dipodomys ordi.—The baculum is similar in every respect to that of *heermanni* (Pl. XII, g).

Measurements of 18 specimens from Montana (1), Wyoming (1), Nebraska (1), Oklahoma (2), Texas (5), Arizona (2), New Mexico (5), and Querétaro (1) are: length, 10.7–12.5 (11.3); height of base, 1.6–2.3 (2.0); width of base, 1.4–2.1 (1.66).

Dipodomys agilis.—The bone in this species averages slightly smaller than others with the large, bulbous basal ends. Otherwise, I can see no differences (Pl. XII, e).

Measurements of eight specimens from California are: length, 9.0–10.3 (9.7); height of base, 1.3–1.9 (1.6); width of base, 1.1–1.4 (1.3).

Dipodomys microps.—Unless the four specimens available are all young, the baculum is the smallest and most delicate in the genus. The basal end is only slightly enlarged and is nearly circular in cross section.

Measurements of four specimens from California (2) and Nevada (2) are: length, 7.3–8.3 (7.85); height of base, 0.8–1.0 (0.9); width of base, 0.6–0.9 (0.75).

Dipodomys deserti.—Although this is one of the larger kangaroo rats, the baculum is relatively and actually (except for *microps*) the smallest of those before me (Pl. XII, a). The basal end is slightly enlarged and somewhat higher than wide; it tapers gradually into the shaft, and the distal end is turned up at nearly a right angle. One specimen has been broken and repaired about a third of the way back from the distal end.

Measurements of four specimens from Nevada are: length, 8.4–9.9 (9.25); height of base, 0.9–1.3 (1.15); width of base, 0.8–1.0 (0.9).

Dipodomys merriami.—The baculum in this species is relatively long. The moderately enlarged basal end tapers gradually into the shaft which has a slight ventral curvature then a dorsal one near the distal end. The tip does not curve dorsally at as acute an angle as in the species previously discussed (Pl. XII, c). The species may be recognized from the bone alone, if it is an adult. The basal end is slightly, if any, higher than wide.

Measurements of 51 specimens from California (22), Nevada (12), Arizona (11), Sonora (3), Tamaulipas (1), and Texas (2) are: length, 9.3–11.7 (10.8); height of base, 1.0–1.7 (1.2); width of base, 0.8–1.3 (1.06).

Dipodomys nitratoides.—The bone in this species is similar to that in *merriami*, but longer. There is no overlap in measurements of length between *nitratoides* and *merriami*. The basal end is about the same size as that in *merriami*. By length and general proportions, the species may be identified from the adult bone alone. Its relationships are definitely with *merriami*, as indicated by the baculum (Pl. XII, d).

Measurements of seven specimens from California are: length, 13.0–13.7 (13.3); height of base, 1.0–1.2 (1.13); width of base, 0.9–1.0 (0.94).

FAMILY CRICETIDAE

In this family there are two general types of bacula. One, found in most of the genera now considered to be in the subfamily Cricetinae, consists of an osseous element with a single cartilaginous tip. The bone varies considerably in shape and size, but is uniform in its simplicity. The other type consists of a short shaft, usually expanded on the basal end, with three finger-like processes (either cartilaginous or osseous) attached to the distal end. This type is found in the Microtinae and in the following members of the Cricetinae: the Old World *Cricetus* and its allies and the New World genera *Neacomys, Nectomys, Oryzomys, Oecomys, Sigmodon, Akodon, Phyllotis,* and *Hesperomys.*

Lawrence (1941) described and figured the bone in *Neacomys,* showing the three finger-like processes on the distal end. Hamilton (1946) pointed out the similarity of the bacula of *Oryzomys* and *Sigmodon* to those of the Microtinae, and Callery (1951) showed the same for the golden hamster, *Mesocricetus (Cricetus) auratus.*

Rinker (1954) presented myological evidence that strongly indicates close relationship between *Orzyomys* and *Sigmodon,* but distant relationships between these two genera and the couplet *Peromyscus* and *Neotoma*. The latter two genera are apparently fairly closely related. Hooper (1959) indicates that the subfamily Cricetinae as now constituted may be polyphyletic. The bacula support this thesis.

GENUS *Reithrodontomys*

The bacula in members of the genus *Reithrodontomys* are rather uniform in shape and size. The bone is a simple curved rod with a dorsoventrally flattened, laterally expanded base that usually is somewhat concave dorsally. The wide base tapers quickly into the shaft which is usually of uniform thickness to near the tip. Relationship with the genus *Peromyscus* is clearly indicated. In fact, the bones of the two genera are often indistinguishable. It is my opinion that within this genus the baculum is of minor taxonomic significance.

Reithrodontomys humulis.—One of the smallest (except for *montanus*) in the genus. The broad basal end is somewhat thicker than the slightly curved shaft (Pl. XIII, c, d), and is concave above. Blair (1942) briefly described and figured the bone for this species.

Measurements of nine specimens from Kentucky (1), West Virginia (1), and Virginia (7) are: length, 5.2–6.4 (6.0); width of base, 0.7–1.0 (0.84).

Reithrodontomys montanus.—The bone in this species is the smallest of any examined. It is very close to that of *humulis* in general shape (Pl. XIII, a, b).

Measurements of two specimens from Nebraska (1) and Oklahoma (1) are: length, 5.2, 5.2; width of base, 0.8, 0.9.

Reithrodontomys burti.—I have seen no specimens. The following is from Benson (1939:149).

"In the shape of the baculum (see figure) *burti* is strikingly distinct from *montanus, megalotis,* and *fulvescens,* in which this bone is distinctly curved and tends to be broader and more flattened at the base. So distinctive is this character that by it alone *burti* can be distinguished from the other harvest mice occurring in Sonora."

Measurements, from Benson (*op. cit.*:148), are: "length of baculum (10 specimens), 7.3 (6.0–8.3)."

Remarks.—From Benson's figure (p. 150) I judge that the bone has a slight upward curvature from the base then straightens out toward the distal end to give it a thin S-curve when viewed laterally.

Reithrodontomys megalotis.—The bone is definitely curved upward throughout its length. The base is broadest in the middle and tapers both fore and aft. Somewhat thicker than the shaft, it has a slight depression on the upper surface. The shaft is nearly uniform in thickness to near the distal end (Pl. XV, g–j). There is some variation, chiefly in size, throughout the vast range of the species. Some of this is shown in the illustrations. In Plate XV, g and h are of a specimen from Washington while i and j are of one from México. The Mexican examples average slightly larger than those from the United States.

Measurements of 34 specimens from Washington (9), Montana (5), North Dakota (5), South Dakota (4), Nevada (5), California (4), Arizona (1), and Texas (1) are: length, 6.2–7.7 (7.16); width of base, 0.5–1.1 (0.88). The same for 29 specimens from México are: length, 7.0–9.2 (7.9); width of base, 0.8–1.1 (.96).

Reithrodontomys sumichrasti.—The basal portion of the bone is scarcely thicker than the shaft; it curves downward proximally. From dorsal view, the broadened base tapers into the shaft more gradually than in most other species and its dorsal depression is barely perceivable. The upcurved shaft is as thick (or thicker) in the middle as near the base. The terminal one-third tapers gradually to a blunt point (Pl. XIII, k, l). The specimen illustrated has a narrow base; in some, the base is nearly twice as wide.

Measurements of 33 specimens from Guatemala (6), Chiapas (9), Veracruz (1), Puebla (1), Michoacán (7), and Jalisco (9) are: length, 7.3–8.7 (8.2); width of base, 0.6–1.2 (0.96).

Reithrodontomys chrysopsis.—The bone is slightly curved, mostly on the terminal one-third. The broad base is definitely thicker than the shaft, tapers rather abruptly into the shaft, and has a definite depression on the dorsal surface. The shaft is of about even thickness to near the blunt tip (Pl. XIII, g, h).

Measurements of ten specimens from México are: length, 6.3–7.4 (7.0); width of base, 0.8–1.0 (0.87).

Reithrodontomys fulvescens.—Bacula in this wide-ranging species are fairly uniform in general configuration. The basal portion is definitely thicker than the shaft and the broad part is depressed above. The upward curvature of the shaft and the width of the basal end vary somewhat, as shown in Plate XIV. Most of this variation seems to be individual. In Plate XIV, a and b and k and l are of specimens from the same population in Michoacán. The only trend that I can detect is that those from Tamaulipas and Arizona appear to be slightly larger than those from farther south in México.

Measurements of eleven specimens from Tamaulipas (8), Nuevo León (1) and Arizona (2) are: length, 7.6–9.4 (8.4); width of base, 0.9–1.2 (1.0). Thirty-six specimens from Chiapas (1), Veracruz (1), Nayarit (4), Jalisco

(9), Oaxaca (2), Guerrero (2), Hidalgo (2), Puebla (1), and Michoacán (14) are: length, 6.9–8.6 (7.7); width of base, 0.8–1.2 (0.94).

Reithrodontomys gracilis.—The bone in this species differs from most others in that the base is about twice as thick as the shaft; it has a deep median depression on the dorsal surface and is sculptured on the edges. The shaft is relatively straight for the first two-thirds, then curves gracefully upwards (Pl. XV, e, f).

Measurements of two specimens from Chiapas are: length, 8.5, 9.0; width of base, 1.1, 1.1.

Reithrodontomys mexicanus.—The base of the baculum is about equal to the shaft in thickness; a slight depression is apparent on its dorsal surface. The shaft is relatively slender and long, and bends upward near the middle (Pl. XIII, i, j).

Measurements of six specimens from Costa Rica (1), El Salvador (1), and Chiapas (4) are: length, 8.3–9.8 (9.1); width of base, 0.8–1.1 (1.0).

Reithrodontomys microdon.—Represented by one adult from Distrito Federal, México, this apparently is the largest of the bacula in the genus (Pl. XV, k, l). Its broad base is thicker than the shaft and has a slight median depression.

Measurements of the one specimen are: length, 9.8; width of base, 1.0.

Reithrodontomys creper.—One specimen represents this species. It differs from others in having a bulbous base and a relatively thick shaft (Pl. XIII, e, f). The one specimen may not be representative, but I do not expect normal variation to account for its configuration. It is one of the smaller bones in the genus.

Measurements of the one specimen are: length, 6.8; width of base, 0.7.

GENUS *Peromyscus*

The bacula in most members of the genus *Peromyscus* may be distinguished from those of other genera, except *Reithrodontomys*. Usually, they are simple rod-like structures with dorsoventrally flattened and laterally expanded basal portions that taper rather abruptly into the slender shafts. The shaft usually curves gracefully upward and ends in a blunt point (sometimes slightly expanded). The expanded base usually has a depression on the dorsal side and in some on the ventral side.

The taxonomic value is questionable in some of the species, but, as pointed out by Blair (1942) some of the subgenera may be distinguished by this element alone. Thus, the subgenera *Peromyscus, Haplomylomys, Podomys,* and *Ochrotomys* may be distinguished if we remove some exceptions such as *banderanus* from the subgenus *Peromyscus* (see p. 57). Hooper (1958) has suggested that *Ochrotomys* be given generic rank.

Peromyscus eremicus.—In this species, the bone is relatively short and thick. The base is broadly expanded, concave above and below, and flattened dorsoventrally as in all members of the subgenus (Pl. XVI, k, l). There is little variation throughout the range of the species except for the population at the tip of Baja California and the one on Carmen Island. These have been considered members of *eremicus* (*P. eremicus eva* and *carmeni*), however, the bacula are radically different from those of eremicus proper (Pl. XVI, i, j), and I doubt that they belong in this species. In the Cabo San Lucas and Carmen Island populations, the bone is small, nearly straight, and has the basal end but slightly expanded. Specimens from north of the cape in Baja California are typical *eremicus*—it is only the cape and Carmen Island populations that are different.

Measurements of three specimens from Cabo San Lucas are: length, 7.9–8.3 (8.1); width of base, 0.8–0.8 (0.8). One specimen from Carmen Island measures 8.0 and 0.7. Thirty specimens from Baja California (5), California (3), Sonora (9), Arizona (6), Nevada (1), and Tamaulipas (6) measure: length, 6.8–9.2 (8.1); width of base, 1.4–2.1 (1.75).

Peromyscus californicus.—This species has the largest of the *eremicus* type baculum. Although not typical, it fits into this group. The broad flat base, which is widest near the middle, has a depression above and below. The base is scarcely higher than the shaft is thick (Pl. XVII, k, l). It differs from typical *eremicus* in the more gradual taper of the base into the shaft and in the taper of the posterior part of the base.

Measurements of 14 specimens from California are: length, 12.6–15.4 (13.7); width of base, 1.7–2.7 (2.0).

Peromyscus pembertoni.—The bone in this species is closest to those of the *eremicus*-like mice. It has a broad base with a shallow depression above and below. There is a small notch at the extreme basal end, which may or may not be significant (Pl. XVII, i, j).

Measurements of four specimens from San Pedro Nolasco Island are: length, 10.3–11.3 (10.75); width of base, 1.9–2.2 (2.05).

Peromyscus crinitus.—The baculum in *crinitus* is similar to that in *eremicus*. The broad base is concave above and below (Pl. XVI, a, b). Hooper (1958), on the basis of the soft parts, thought that *crinitus* should be placed in the subgenus *Peromyscus*. This is not indicated by the baculum. I suspect that *crinitus* is one of those intermediates between groups with some characteristics close to one and others close to another group.

Measurements of four specimens from California (1), Arizona (1), and Nevada (2) are: length, 7.5–8.6 (8.2); width of base, 1.5–2.1 (1.8).

Peromyscus collatus.—The baculum is relatively short with a wide base that is concave above and below. It is similar to that of *crinitus,* but smaller. The bone is fairly typical of the *eremicus* group (Pl. XVI, c, d).

Measurements of four specimens from Turner's Island are: length, 5.7–6.6 (6.15); width of base, 1.3–1.7 (1.5).

Peromyscus pseudocrinitus.—Two specimens from Coronados Island are available. The broad, flattened base which is concave above and below is fairly typical of the *eremicus* group (Pl. XVI, e, f).

Measurements of the two specimens are, respectively: length, 8.5, 8.3; width of base, 1.8, 1.8.

Peromyscus guardia.—The baculum of *guardia* from Angel de la Guarda Island is quite like that of *stephani* except that it is shorter and the base is wider, with a notch on the posterior part. It differs from the *eremicus* type in that the ventral surface of the base is not concave (Pl. XVII, e, f). *P. guardia mejiae* from Mejía Island nearby has a much smaller base more like that of *slevini* (Pl. XVII, g, h). The bone of *P. guardia interparietalis* from South San Lorenzo Island is more like that of *pembertoni* than that of *g. guardia.* Its base is concave above and below. Close relationships between *guardia, mejiae,* and *interparietalis* are not indicated by the bacula.

Measurements of three specimens from Angel de la Guarda Island are: length, 9.0–10.8 (9.7), width of base, 1.5–1.8 (1.6). One specimen from Mejía Island measures 11.6 by 1.3. Two specimens from South San Lorenzo Island measure, respectively: 9.5, 8.6; 1.9, 1.6.

Peromyscus stephani.—The bone is similar to that of *slevini,* except that the base is less rounded and slightly smaller (Pl. XVII, c, d). It has a depression above, but is convex below—a characteristic of the subgenus *Peromyscus.* However, it is now placed in the subgenus *Haplomylomys.*

Measurements of four specimens from San Esteban Island are: length, 11.0–12.5 (11.7); width of base, 1.3–1.4 (1.3).

Peromyscus caniceps.—The bone in this species has a slightly expanded base which is convex dorsally and flat on the ventral side. It is not an *eremicus* type bone (Pl. XVI, g, h). In configuration and size it is similar to those from Cabo San Lucas and Carmen Island.

Measurements of three specimens from Monserrate Island are: length, 8.1, 8.3, 8.6; width of base, 0.8, 0.8, 1.0.

Remarks.—Specimens of *P. caniceps, eremicus, carmeni,* and those *eremicus eva* from Cabo San Lucas constitute a group that differs from all other *eremicus*-like *Peromyscus* in characters of the baculum. Further study of the entire animals is indicated to determine their true relationships.

Peromyscus maniculatus.—The baculum in this species is a fairly simple, slightly curved bone with an expanded base that has a concavity on top. Variation is primarily in length of bone and in width of the basal end (Pl. XVIII). *P. m. bairdi* and *nubiterrae* have the shortest bones whereas *gracilis* has the longest. *P. m. margaritae* from Santa Margarita Island, Baja California, has the widest base. Except for the island races (*margaritae, dubius,* and *exiguus*), northern representatives appear to have longer bacula than the southern races. Measurements are given in Table 2. In general configuration, the bone is similar to that of *leucopus.*

TABLE 2

MEASUREMENTS OF BACULA OF *Peromyscus maniculatus* (MM.)

Locality and subspecies	Number	Length		Width of Base	
		Range	Mean	Range	Mean
Indiana and Michigan, *bairdi*	9	6.9–7.6	7.2	1.0–1.2	1.1
Michigan, *gracilis*	22	8.7–9.8	9.3	1.0–1.5	1.2
Washington, *oreas*	5	8.3–9.4	8.7	1.1–1.2	1.16
Washington, *gambeli*	10	8.0–8.9	8.5	1.0–1.4	1.2
Oregon, *rubidus*	5	8.4–9.5	8.7	1.1–1.3	1.2
S. Dak., Wyo., Montana, *osgoodi*	33	7.2–9.0	8.1	0.9–1.4	1.15
New Mexico, *rufinus*	11	7.5–8.6	7.8	1.0–1.3	1.1
North Carolina, *nubiterrae*	13	6.9–7.7	7.2	1.0–1.3	1.1
Michoacán, D. F., Jalisco, Nayarit, *labecula*	34	6.9–8.4	7.6	1.0–1.5	1.2
Margarita Island, *margaritae*	2	8.4–8.4	8.4	1.5–1.9	1.7
Todos Santos Island, *dubius*	3	8.6–9.5	8.9	1.1–1.2	1.17
San Martin Island, *exiguus*	9	8.2–9.2	8.8	1.1–1.7	1.4
Total	156				

Peromyscus sejugis.—Four adult bones from Santa Cruz Island and one from San Diego Island, Gulf of California, México, are available. They are similar to those of small *maniculatus*. The proximal expansion is moderate, concave dorsally, convex ventrally, and thinner than the middle of the shaft (Pl. XIX, e, f).

Measurements of the four specimens from Santa Cruz Island are: length, 9.5–9.7 (9.6); width of base, 1.0–1.2 (1.05). The one specimen from San Diego Island measures: length, 8.4; width of base, 0.9.

Peromyscus slevini.—The baculum has a bulbous base, somewhat rounded, concave above and convex below. The base is appreciably thicker in dorsoventral diameter than the shaft (Pl. XVII, a, b). Relationships with members of the subgenus *Peromyscus* are indicated.

Measurements of 16 specimens from Santa Catalina Island, Baja California, are: length, 10.5–13.1 (11.7); width of base, 1.3–1.6 (1.4).

Peromyscus melanotis.—The baculum is relatively small and the base is moderately expanded (Pl. XIX, c, d). It is similar to the smaller *maniculatus.* Twenty specimens show fair uniformity in size, but two from Jalisco are short with wide bases (length, 7.8, 7.3; width of base, 1.2, 1.2). They appear to be adults. These two are not included in the measurements given below. One specimen from Distrito Federal had been broken and mended near the middle of the shaft.

Measurements of 20 specimens from Veracruz (7), México (3), Distrito Federal (7), Morelos (1), and Michoacán (2) are: length, 7.9–8.8 (8.4); width of base, 1.0–1.3 (1.1).

Peromyscus leucopus.—The bone is slightly curved, and expanded on the proximal end, which is concave above and convex below. The angle at which the base slopes into the shaft varies somewhat, usually fairly gradually as shown in Plate XIX, h, but often more abruptly. The bone may terminate distally in a blunt point or in a slight expansion. In some old specimens, the cartilaginous extension (cap) appears to be ossified. One specimen from Michigan had been broken in the middle of the shaft and repaired. Another from Oaxaca appears to have been broken near the distal end.

Some geographic variation in size is indicated. Ten specimens from Tamaulipas measure largest: length, 9.0–10.6 (10.0); width of base, 1.3–1.6 (1.4). Eight specimens from Oaxaca are smallest with the following measurements: length, 8.3–9.6 (9.0); width of base, 1.3–1.4 (1.32). Thirty specimens from Michigan are intermediate with the following measurements: length, 8.5–10.9 (9.4); width of base, 1.1–1.5 (1.3). The entire series of 89 specimens from Oaxaca (8), Veracruz (3), Puebla (6), San Luis Potosí (1), Quintana Roo (2), Campeche (3), Neuvo León (1), Tamaulipas (10), Tennessee (2), Kentucky (2), Virginia (3), West Virginia (4), Pennsylvania (3), Ohio (2), Michigan (30), Minnesota (5), Iowa (1), Indiana (1), Illinois (1), and Missouri (1) are: length, 8.1–10.9 (9.5); width of base, 1.0–1.6 (1.3).

Remarks.—Sprague (1939) pointed out the difference in size between *maniculatus* and *leucopus* of Central United States. When we consider *maniculatus* throughout its geographic range, these differences disappear. *P. m. gracilis* (Table 2), for instance, is well within the size range of *leucopus.* Blair (1942) indicated that there were no qualitative differences between the two species. With this I concur.

Peromyscus gossypinus.—The baculum may be nearly straight or slightly curved. The base is relatively wide for the length of the bone, slightly concave above and convex below (Pl. XIX, i, j). It is longer and broader than the bone in *leucopus*.

Measurements of two bones from Florida are, respectively: length, 11.5, 10.2; width of base, 1.5, 1.3.

Peromyscus boylei.—The bone is of medium size for the genus, and the base is not greatly expanded (Pl. XIX, k, l). There is individual as well as geographic variation in size and proportions. Thirteen specimens from Chiapas average 11.1 mm. in length, and 16 from Nayarit average 12.6 mm. The average width of the base is 1.4 mm. for Chiapas and 1.3 for Nayarit. Six specimens from Texas (3) and Arizona (3) average 13.0 by 1.3 for the same dimensions. There is an apparent reduction in length as one goes from north to south. Two specimens, one from Querétaro and one from Nayarit are abnormal in that the distal ends are rather sharply upturned. The distal end is capped with a small, rounded mass of cartilage (Clark, 1953: Hooper, 1958).

Measurements of 69 adult specimens from Texas (3), Arizona (3), New Mexico (1), California (1), Distrito Federal (1), Hidalgo (1), Puebla (1), Querétaro (1), Nayarit (16), Tamaulipas (3), Jalisco (5), Chiapas (13), Veracruz (4), Michoacán (13), Oaxaca (1), and San Pedro Nolasco Island, Baja California (2) are: length, 10.0–14.2 (11.8); width of base, 1.1–1.6 (1.3).

Peromyscus perfulvus.—The baculum in this species is peculiar in that it has a double S-curve in many cases (six of thirteen). The one figured (Pl. XXI, o, p) has an indication of this toward the tip. The bulbous base is relatively high for its width, and has a definite concavity on the dorsal side; in some specimens the basal portion is U-shaped in cross section. Although definitely of the *Peromyscus* type, the configuration of the bone is unique in the group.

Measurements of three specimens from Michoacán are, respectively: length, 12.3, 13.7, 14.0; width of base, 1.4, 1.4, 1.4.

Peromyscus hylocetes.—Base of baculum expanded laterally, flattened dorsoventrally, and with a slight depression on its dorsal side. Shaft thin and slightly curved (Pl. XXI, m, n). There is one deformed specimen; near the middle, the shaft is bent to the right at a nearly 90 degree angle.

Measurements of 24 specimens from Michoacán (20) and Jalisco (4) are: length, 10.0–12.0 (11.0); width of base, 1.0–1.5 (1.3).

Peromyscus pectoralis.—The baculum has a relatively broad base that tapers gradually into the shaft, which curves gracefully. Dorsally, the base is concave; its height is scarcely greater than the diameter of the shaft (Pl. XX, g, h). The bone is similar to that of *boylei*, but the cartilaginous cap is distinctly longer (Clark, 1953; Hooper, 1958).

Measurements of eight specimens from Querétaro (6) and Hidalgo (2) are: length, 11.0–12.5 (12.0); width of base, 1.3–1.6 (1.46).

Peromyscus truei.—The baculum is long, slender, slightly curved, and moderately expanded at the base (Pl. XX, a, b).

Measurements of 26 specimens from Distrito Federal (5), Hidalgo (4), Michoacán (2), New Mexico (11), and California (4) are: length, 12.5–15.9 (14.4); width of base, 1.0–1.8 (1.5).

Peromyscus nasutus.—The baculum of *nasutus* is indistinguishable from that of *truei*. In the series measured, it averages slightly longer (Pl. XX, c, d). One specimen has an S-shaped curvature (see also Tamsitt, 1958).

Measurements of 22 specimens from New Mexico are: length, 13.2–16.2 (14.7); width of base, 1.2–1.7 (1.4).

Peromyscus dificilis.—The baculum is long, slender, curved, and the base is moderately expanded laterally. It is similar to those of *nasutus* and *melanophrys* (Pl. XX, k, l), except that the base averages slightly wider.

Measurements of 14 specimens from Nuevo León (1), Hidalgo (6), Distrito Federal (5), and Veracruz (2) are: length, 14.4–18.5 (16.4); width of base, 1.4–1.9 (1.7).

Peromyscus melanophrys.—The bone is long, slender, definitely curved, and has a moderately expanded proximal end which tapers gradually into the shaft (Pl. XX, i, j). The basal part is depressed dorsally. It is similar to that of *nasutus,* but with slightly greater curvature.

Measurements of twelve specimens from Guerrero (3), Puebla (1), Michoacán (3), Hidalgo (1), and Oaxaca (4) are: length, 14.6–17.1 (15.9); width of base, 1.3–1.9 (1.6).

Peromyscus mexicanus.—A long, slender, slightly curved bone with a relatively small rounded basal end, sometimes depressed dorsally (Pl. XX, e, f). Some of the variation in length of the series measured probably is owing to age differences.

Measurements of 15 specimens from Chiapas (9) and Veracruz (6) are: length, 12.2–17.4 (14.7); width of base, 1.1–1.6 (1.3).

Peromyscus yucatanicus.—The baculum is long and slender, and gently curved. The base is wider than high, but expanded relatively little; in some specimens it is nearly round, but usually it is flattened dorsoventrally and depressed dorsally (Pl. XXI, q, r).

Measurements of eleven specimens from Quintana Roo are: length, 13.5–15.3 (14.3); width of base, 1.2–1.6 (1.3).

Peromyscus mekisturus.—One specimen is available, and it appears to be an adult. It is a short bone (length, 7.2; width of base, 1.1), slightly curved, and the expanded basal part is relatively deep with a concavity on the ventral surface (Pl. XIX, a, b). It is quite different from the bone in *melano-phrys* (Pl. XX, i, j), in which group it is now placed. More material is needed before any definite conclusions can be reached regarding relationships.

Peromyscus banderanus.—Except for *nuttalli,* this species has the smallest baculum of any *Peromyscus* examined. It is a simple, slightly curved rod with a rounded, bulbous base which is occasionally wider than high. From the expanded base the bone tapers to the pointed tip (in a few there is an indication of a swelling on the tip). The dorsal part of the base has a slight concavity (Pl. XXI, i, j). As Hooper (1958) pointed out, *banderanus* stands apart from other *Peromyscus* in the phallus as well as the baculum.

Measurements of 13 specimens from Guerrero (4), Michoacán (2), Nayarit (3), and Jalisco (4) are: length, 3.4–4.8 (4.1); width of base, 0.5–0.7 (0.55).

Peromyscus floridanus.—The bone is a simple, slightly curved rod with a rounded, bulbous base barely wider than high. From this base, the bone tapers first rather rapidly then gradually to the tip (Pl. XXI, k, l). It is small relative to the size of the animal.

Measurements of six specimens from Florida are: length, 5.7–8.2 (7.4); width of base, 0.7–0.7 (0.7).

Peromyscus (Ochrotomys) nuttalli.—The baculum in this species does not conform to the ordinary *Peromyscus* pattern. The laterally expanded base tapers into the shaft which continues to the tip as a parallel-sided bone; the tip is blunt, not pointed (Pl. XXI, g, h), and may be slightly expanded. One specimen has a median projection, backward, and two winglike projections laterally on the basal portion; all others are rounded, but dorsoventrally flattened. *P. nuttalli* has the smallest bone of any *Peromyscus* examined. Hooper (1958) suggested generic rank for this species.

Measurements of two specimens from Alabama (1) and Virginia (1) are, respectively: length, 3.8, 3.4; width of base, 1.0, 0.8.

GENUS *Neotomodon*

Neotomodon alstoni.—The baculum in this species is like that of a small *Peromyscus.* It is a simple bone with a smoothly bulbous base (wider than high) that tapers abruptly into the shaft, which curves gently upward to the pointed tip (Pl. XXI, e, f).

Measurements of seven specimens from Distrito Federal (2), Veracruz (1), Morelos (3), and Michoacán (1) are: length, 4.6–5.3 (5.1); width of base, 0.6–0.9 (0.7).

GENUS *Onychomys*

The baculum in this genus is distinct from all others examined in the family Cricetidae. It is a short bone with a round knob for a base, and with a laterally flattened shaft (Pl. XV). Relationships with other members of the family are obscure.

Onychomys leucogaster.—The baculum is as described above and, in addition, has a shaft that, in some, curves gently upward (Pl. XV, a, b). In this way it differs from *torridus*. Also, the shaft is more blade-like than in the latter. The bone was described and figured by Blair (1942).

Measurements of ten specimens from Oklahoma (2), Texas (1), New Mexico (3), Wyoming (1), and Nebraska (3) are: length, 4.2–5.0 (4.7); width of base, 0.5–1.0 (0.8).

Onychomys torridus.—The bone differs from that of *leucogaster* in having a relatively straight, less blade-like shaft; otherwise similar (Pl. XV, c, d).

Measurements of eleven specimens from California (2), Arizona (2), and Nevada (7) are: length, 4.5–5.3 (4.7); width of base, 0.5–0.7 (0.66).

GENUS *Baiomys*

Baiomys musculus.—The baculum in *Baiomys* is peculiar to the genus, and has no resemblance to those found in *Peromyscus,* in which genus it has been placed in the past (see Blair, 1942). The bone is small, the basal part is convex dorsally, concave ventrally to form a longitudinal groove. Occasinally the groove extends nearly the length of the bone (Pl. XXI, b), but mostly it does not extend beyond the widened basal part. The distal end may be bifid or it may be a knob-like swelling. There is considerable variation in detail of the outline as seen from above or below. Some of this is shown in Plate XXI, a–d. *Baiomys* stands apart, in the cricetids examined, in regard to bacular characters. It should, I believe, be retained as a separate genus.

Measurements of 26 specimens from Chiapas (9), Michoacán (2), Guerrero (2), Jalisco (2), and Oaxaca (11) are: length of shaft, 3.0–3.8 (3.3); width of base, 0.6–1.2 (0.9).

Baiomys taylori.—Material representing this species is not too satisfactory. The baculum is of the same general shape as that of *musculus,* but the basal end is less expanded and the middle of the shaft is thinner; this accentuates the enlarged distal end. They are close to the *musculus* figured on Plate XXI, c, d. Blair (1942) described and figured, in outline, the bone of this species.

Measurements of six specimens from Distrito Federal (1), Jalisco (1), and Michoacán (4) are: length, 2.9–3.2 (3.1); width of base, 0.6–0.9 (0.7).

GENUS *Neotoma*

In a previous paper (Burt and Barkalow, 1942), the bacula of ten species were described, and indicated relationships were discussed. Two additional kinds, *torquata* and *ferruginea* (placed in species *mexicana* by Hall and Kelson, 1959), are described and figured here. One of the species included in the previous paper (*magister*) has since been synonymized with *floridana* (Schwartz and Odum, 1957). There is considerable diversity in size and configuration of the bone in the various species. The differences will be pointed out in the separate accounts below. It is my opinion that the bone is useful in resolving questionable relationships of different species in this genus. This was apparent in the *floridana-magister* complex. Geographic variation, particularly in size, is displayed in the *lepida* group, the only one in which we have a sufficiently large series from widely spread localities. This group (*lepida*) should be placed at the beginning or end of the linear arrangement, not in the middle. I choose, arbitrarily, to place it at the beginning. *N. stephensi* is considered to be a distinct species by Hall and Kelson (1959) and by Hoffmeister and de la Torre (1959).

Neotoma lepida.—The baculum in this group differs from those of all others in that the shaft is long (10 mm. or more in adults), slender, and definitely curved upward; the basal end is small (2.6 mm. or less in width) and usually dumbbell shaped in cross section (Pl. XXII, k, l). The series at hand is not large enough to mean very much, but there is an indication of geographic variation in size. Specimens from the mainland and certain islands of Lower California average largest and those from San Bernardino County, California, smallest of the entire series. Age variation is also part of the picture, but we have no specimens of known age. One specimen from Angel de la Guarda Island appears to be abnormal. It measures only 8.4 mm. in length, the shaft is but slightly curved, and the distal end is definitely enlarged. Additional material is necessary before definite conclusions can be formulated on the character and relationships indicated by this one bone. For the others, measurements are given for each group from a distinct locality.

California, San Bernardino County (3): length, 9.9–11.5 (10.6); width of base, 1.6–1.9 (1.8). Riverside County (4): length, 14.9–17.7 (16.3); width of base, 2.0–2.4 (2.3). *Nevada,* Clark County (8): length, 10.0–12.7 (11.1); width of base, 1.6–2.2 (1.9). *Arizona,* Yuma County (4): length, 13.2–15.1 (14.3); width of base, 1.7–2.0 (1.9). *Lower California,* Magdalena Id. (2),

Santa Margarita Id. (6), San Marcos Id. (3), mainland (6); total (17): length, 14.0–20.9 (17.3); width of base, 1.5–2.6 (2.2).

Neotoma bunkeri.—The bone in this island species is the largest of any woodrats examined. Its shape is similar to that of *lepida*, its closest relative (Burt and Barkalow, 1942:290). One bone from Coronados Island, Gulf of California, measures: length, 21.83; width of base, 2.6.

Neotoma floridana.—The baculum in this species (now includes *magister*) is relatively short with a wide base and a slender, slightly curved shaft that turns up on the distal end. In cross section, the base is deeply concave ventrally, slightly concave dorsally (Pl. XXII, e, f).

Measurements of eleven specimens from Kansas (8) and Kentucky (3) are: length, 6.6–7.7 (7.2); width of base, 2.6–4.1 (3.1).

Neotoma micropus.—The bone in this species cannot be distinguished with certainty from that of *floridana*. Close relationships are indicated (Pl. XXII, g, h). It is interesting that the geographic ranges of *floridana* and *micropus* abut one another from southern Texas to Kansas, but they do not overlap.

Measurements of five specimens from Oklahoma (1), Texas (3), and Tamaulipas (1) are: length, 5.8–7.7 (6.5); width of base, 2.5–3.3 (3.0).

Neotoma albigula.—The bone in this species differs from those of *floridana* and *micropus* in having a small round knob on the distal end, instead of an upturned point. Otherwise, the bones are quite similar in the three species. I think there is no doubt of the close relationship here (Pl. XXII, i, j).

Measurements of 13 specimens from Sonora (2), Arizona (4), New Mexico (6), and Texas (1) are: length, 5.9–7.4 (6.4); width of base, 2.6–3.4 (2.9).

Neotoma mexicana.—The bone in mexicana (Pl. XXII, c, d) has a slightly enlarged proximal end which tapers abruptly into a dorsoventrally flattened shaft. The shaft then either tapers gradually toward the tip or (usually) broadens near the center and then tapers again toward the tip which, in turn, expands into a slight knob. In cross section, the base may be partially dumbbell-shaped, or it may be concave ventrally and nearly flat dorsally.

Measurements of 21 specimens from Arizona (2), New Mexico (17), and Texas (2) are: length, 5.2–6.9 (6.0); width of base, 1.5–2.6 (2.1).

Neotoma stephensi.—The bone in this species was recently described and figured by Hoffmeister and de la Torre (1959). In the adult form the bone is wedge shaped, tapering gradually from the expanded base to the tip, which may or may not be slightly bulbous. The bone is dorsoventrally flattened for its basal four-fifths. From an end view the base is nearly flat on one surface (dorsal?) and slightly concave on the other. It is very close in configuration to some specimens of *mexicana*. Hoffmeister and de la Torre (*ibid.*) concluded that, based on the baculum, *stephensi* would be "nearest to *N. mexicana* and *N. phenax*." I would place it nearest *mexicana*. The above authors give 5.0 mm. for the length of their longest bone. One bone (UIMNH 18682) that I measured is 4.7 mm. long and the base is 1.6 mm. wide. Thanks to Hoffmeister, I have been able to examine his material.

Neotoma torquata.—The baculum in this species is shaped like a violin, when viewed from above or below. The laterally expanded basal part constitutes about one-half the length of the bone. The short, rounded, distal part is slightly curved and the base, in cross section, is somewhat dumbbell-shaped (Pl. XXIII, k, l). It is perhaps closest, in general configuration, to that of *mexicana*, but differs in certain respects. Hall and Kelson (1959) consider *torquata* a subspecies of *mexicana*.

Measurements of three specimens from Puebla are: length, 6.9–7.3 (7.1); width of base, 1.9–2.5 (2.2).

Neotoma ferruginea.—The bone in this species is unlike any others that I have examined. The base, which is wider than high, has a rounded proximal end then a small projection, or shoulder, on either side. It then tapers slightly, distally, broadens briefly, and finally tapers into the small, short shaft. In addition to the above is a thin shelf of bone on either side that extends laterally beyond the main part of the base (Pl. XXIII, c, d). Close relationship with *mexicana* (Hall and Kelson, 1959) is not indicated by the baculum.

Measurements of two specimens from Jalisco are, respectively: length, 4.0, 5.2; width of base, 0.8, 1.1.

Neotoma alleni.—The baculum in *alleni* resembles that in *albigula* in that the distal, rounded shaft is slightly curved and terminates in a small knob. It resembles *floridana* in the general shape of the proximal end as viewed from above or below. In cross-sectional aspect, the base is not as deep, nor is the concavity as pronounced as in *floridana* (Pl. XXIII, i, j). Possible relationships with *floridana* were pointed out previously (Burt and Barkalow, 1942:295).

Measurements of two specimens from Jalisco are: length, 6.0, 6.8; width of base, 2.4, 3.0.

Neotoma fuscipes.—The baculum in this species differs from all others in that the base is relatively much wider, and is definitely dumbbell-shaped in cross section. The shaft is short and blunt (Pl. XXII, a, b).

Measurements of two specimens from California are: length, 5.0, 5.5; width of base, 3.5, 3.6.

Neotoma cinerea.—The baculum in *cinerea* resembles that of *mexicana* superficially only in the slight bulge in the middle of the shaft (Pl. XXIII, m, n). "The distal end is slightly expanded. The basal portion is quadrate in cross section with a slight concavity on both dorsal and ventral surfaces. In this respect the bone differs from those of all other *Neotoma*." (Burt and Barkalow, 1942:296).

Measurements of eight specimens from South Dakota (1), Wyoming (2), Montana (3), and Washington (2) are: length, 4.8–6.0 (5.3); width of base, 1.3–1.9 (1.5).

Neotoma (Teanopus) phenax.—This species possesses the smallest baculum of any *Neotoma*, if the one specimen available is not abnormal. The dorsoventrally flattened bone has no distinct shaft as do the others (Pl. XXIII, a, b). The one bone from Sonora measures: length, 3.6; width of base, 1.5.

GENUS *Ototylomys*

Ototylomys phyllotis.—The baculum in this species is similar to those of some *Neotoma*, especially in the basal region. In cross section, the base is dumbbell-shaped with the greater concavity dorsally. From the base the bone tapers fairly rapidly for nearly one-half its length, then more slowly to near the tip where there is again an expansion to form a small knob on the distal end, which is slightly upturned (Pl. XXIII, e, f).

Measurements of three specimens are: length, 5.0, 6.0, 6.3; width of base, 1.5, 1.7, 2.7.

Remarks.—Relationships, as indicated by the baculum, are with the woodrats (*Neotoma*). In a linear arrangement, I would place them close to the latter genus. I do not know whether the three finger-like projections characteristic of microtines and oryzomines are present. I suspect not, because they are not present in the closely related genus *Tylomys*, of which we have one cleared and stained specimen.

GENUS *Tylomys*

Tylomys nudicaudus.—The baculum in the one specimen available from Chiapas is a simple bone with an enlarged basal end, broader than high, that tapers rather abruptly into the shaft which ends in a blunt point. It is woodrat-like in general character. It does not have the three distal finger-like structures found in *Oryzomys*. It is much smaller and simpler than the baculum of *Ototylomys*.

Measurements are: length, 2.8; width of base, 0.6.

GENUS *Nyctomys*

Nyctomys sumichrasti.—The baculum in *N. sumichrasti* is quite similar to that found in *Neotoma albigula*. The base is relatively less expanded and the bone is more delicate throughout. The base is, in cross section, some-what dumbbell-shaped with the greater concavity ventrally. The base tapers abruptly into the slender shaft which curves gently upward and terminates in a small, rounded knob (Pl. XXIII, g, h).

Measurements of one specimen from Oaxaca are: length, 5.7; width of base, 1.5.

Remarks.—Definite relationships with the woodrats (*Neotoma*) are indi-cated by the structure of the baculum. We have no cleared and stained material, but if it is eventually determined that the three finger-like pro-cesses on the distal end are absent the evidence for relationships with *Neo-toma* rather than with *Oryzomys* will be strengthened.

GENUS *Nectomys*

Nectomys squamipes.—The shaft of the baculum has a relatively broad base with two condyle-like structures which are concave and, when viewed from the end, form a dumbbell-shaped basal end. Just anterior to the wide base, the shaft flattens dorsoventrally and tapers into a laterally flat-tened tip. The shaft is slightly curved (Pl. XXIII, o, p). There are three finger-like, cartilaginous processes on the distal end (not shown in the illustration).

One specimen from British Guiana measures 5.7 mm. in length of shaft and 3.4 in width of base.

Remarks.—Affinities with *Oryzomys* are indicated by the baculum. This agrees with the suggestion by Hershkovitz (1944), based on cranial charac-ters.

GENUS *Neacomys*

I have no specimens of this genus, but from the short description and the figures given by Lawrence (1941) it appears to be similar to some of the *Oryzomys*. The shaft is expanded both laterally and dorsoventrally at the base, and there are three cartilaginous processes on the distal end. No meas-urements were given.

GENUS *Oryzomys*

Oryzomys couesi.—The shaft of the baculum is similar to those of the microtines. The base is broad and dumbbell-shaped in cross section. It tapers rapidly into the laterally compressed shaft that has a keel along its ventral surface and a slight knob on the distal end. Anteriorly is a thin median and two rather heavy cartilaginous processes (Pl. XXIV, i, j).

Measurements of 12 specimens from México are: length of shaft, 3.0–4.5 (3.6); width of base, 1.7–3.0 (2.2).

Oryzomys palustris.—One dry specimen from Alabama and three in liquid from Kentucky resemble the bones of *O. couesi* in every respect. The one dry specimen (shaft) is 3.5 mm. long and the base is 2.2 mm. wide.

Oryzomys melanotis.—The baculum in this species is more delicate than that of *couesi*. One specimen from Guerrero has a ventral keel on the shaft, the others have no keel. I suspect the latter are all of immature animals. The basal end of the shaft is dumbbell-shaped in cross section. The general outline of the bone is similar to that of *couesi* (Pl. XXIV, e, f). However, the median cartilaginous process on the distal end is little more than a spicule. The distal end of the shaft is slightly expanded into a small knob.

Measurements of seven (immatures?) from Jalisco and one adult from Guerrero are: length of shaft, 3.1–3.9 (3.5); width of base, 1.2–2.0 (1.5).

Oryzomys alfaroi.—We have two specimens from Puebla (1) and Chiapas (1), both in liquid. The base of the shaft is broad as in the microtines. In cross section it is dumbbell in shape in one specimen, more concave dorsally than ventrally in the other (Pl. XXIV, a, b). The shaft is gently curved upward and slightly enlarged on the distal end. Attached to the distal end are the three cartilaginous processes; the central one is smallest, the lateral ones curve outward then inward. The relationships with microtines and with *Sigmodon* are, I think, apparent.

Oryzomys fulvescens.—The bone in this species is similar to that of *alfaroi* in general configuration. It differs from that species in having a distinct keel on the ventral part of the shaft and in having a relatively thicker median cartilaginous process anteriorly. I have no dry specimens for measurements.

GENUS *Dicrostonyx*

Dicrostonyx hudsonius.—The base of the shaft of the baculum has a condyle-like swelling on either side with a distinct median notch between them. The inflated basal portion tapers into a parallel-sided shaft that is rounded at the distal end. Three cartilaginous processes project from the terminal portion (Pl. XXV, e). Hamilton (1946:381) indicates that in a single specimen of *D. groenlandicus rubricatus* "the lateral digital processes are partially ossified."

The supposed close relationship with *Lemmus* is not indicated by the baculum. Dearden (1958) gives the following measurements for three specimens of *groenlandicus*: length of shaft, 2.85–3.1 (2.95); width of base, 1.5–1.61 (1.54).

GENUS *Synaptomys*

Synaptomys cooperi.—The rather smoothly rounded base has a shallow median notch posteriorly and a swollen, condyle-like projection on either side. This tapers into the shaft which expands slightly at the terminus. Connected with this by cartilage are three small osseous processes; they are relatively short and thick, and the lateral ones are curved inward (Pl. XXV, c). As suggested by Hamilton (1946:381), the one figured by Hibbard and Rinker (1942) might possibly have lost the lateral processes in preparation. However, it is possible that in the isolated population of *S. c. paludis* there is but one process.

Measurements of seven specimens from Michigan (6) and Virginia (1) are: length of shaft, 1.8–2.5 (2.3); width of base, 0.7–1.2 (1.0).

GENUS *Lemmus*

Lemmus helvolus.–Hamilton (1946:381) described the bone as follows. "The long prominent digitate processes . . . set it off from all other genera of the *Lemmi* group and suggests affinities with the *Microti*. The stalk is comparable in appearance to immature specimens of *Microtus pennsylvanicus* . . . and *Clethrionomys gapperi,* although more delicate in structure. The entire dorsal surface [probably ventral in my interpretation] of the base has a slight convexity, while the ventral area is slightly concave medially, possessing two ill-defined condyles. The digital processes are unique in their large size, the median process approximating four-fifths the length of the stalk. It broadens perceptibly at its base, each lateral margin carrying a spur that partially encloses the distal portion of the stalk. Just above these spurs arise the lateral processes, somewhat shorter and less robust than the median element. Measurements of the single specimen available: overall length, 3.9 mm; stalk, 2.1 mm; width of base, 1.1 mm; length of median process, 1.8 mm; lateral process, 1.3 mm." (Pl. XXV, d, after Hamilton) For length of shaft of two specimens of *L. trimucronatus,* Dearden (1958) gives 2.68 and 2.71.

GENUS *Clethrionomys*

Clethrionomys gapperi.–The dorsoventrally flattened basal part of the baculum is broadly circular as seen from above, with the dorsal surface concave. It tapers rapidly to a blunt tip which is connected by cartilage to three ossified processes. The processes vary in length, and are relatively long for the microtines (Pl. XXV, a, b). I do not see the "spur" on the central process in any of our specimens (Hamilton, 1946:382).

Measurements of eleven specimens from Maine (2), Michigan (6), Minnesota (1), North Carolina (1), and Washington (1) are: length of basal stalk, 2.0–2.7 (2.4); width of base, 1.0–1.7 (1.3).

GENUS *Phenacomys*

Phenacomys longicaudus.—The baculum in this species is quite distinct. The distal end of the shaft is expanded, "forming an ill-defined trilobed extremity," (Hamilton, 1946:381). My specimen does not have the processes ankylosed to the distal end of the shaft as did Hamilton's (*ibid.:* 382). Also, this is the only microtine examined by me in which the lateral osseous processes are larger than the median one. The shaft, from lateral view, curves upward (Pl. XXV, f). Dearden's (1958) specimens had cartilaginous processes with a small osseous nodule in the central one. His measurements for three specimens are: length of shaft, 2.53–2.70 (2.61); width of base, 1.67–1.81 (1.76).

GENUS *Microtus*

Microtus pennsylvanicus.—The basal part of the shaft is broad and angular; the lateral portions are condyle-like. Viewed from the end, the base is dumbbell-shaped with the greater concavity on the dorsal aspect (Pl. XXIV, c, d). The shaft proper tapers to a blunt point which has a cartilaginous connection with the three distal processes. These processes curve upward. In development, the main shaft ossifies first, then the large central process followed closely by the smaller laterals. Hamilton (1946) pointed out the great variation, individual and age, in bacula of this species.

Measurements of 17 specimens from Michigan (13), South Dakota (1), North Dakota (1), and Montana (2) are: length of base (stalk), 2.4–3.1 (2.8); width of base, 1.3–2.1 (1.6). Hamilton's (1946) series of 20 adults averaged slightly larger than the above (stalk, 3.1; width of base, 1.9).

Microtus californicus.—Similar to that of *pennsylvanicus* except that the basal portion is dorsoventrally flattened and not medially notched in the one adult specimen (shaft only) available. The shaft is 3.2 mm. long and 1.7 mm. wide at the base.

Microtus longicaudus.—From the five dry specimens available, the baculum in this species appears to be similar in every respect to that of *M. pennsylvanicus.*

Measurements of two specimens from Montana are: length of shaft, 3.1, 2.8; width of base, 1.8, 1.5.

Microtus mexicanus.—The shaft of the baculum in this species is fairly typical of microtines. From an end view, the broad basal part is deeply notched both dorsally and ventrally to give it a dumbbell appearance. The distal processes are cartilaginous with a large osseous nodule in the central one and a very small osseous nodule on each side of and slightly posterior to the tip of the main shaft (Pl. XXIV, g, h). This is the only microtine examined that has this arrangement. The nodules in *ochrogaster* are more distal. I am not certain that all specimens of *mexicanus* have the lateral nodules in the adult stage. In good preparations they are apparent, in some of the poorer preparations, they seem to be absent.

The shaft of one specimen is 3.0 mm. long and its base is 1.4 mm. wide.

Microtus oeconomus.—The baculum of this species, represented by two specimens in liquid from Kodiak Island, Alaska, differs from *pennsylvanicus* in having a greater amount of cartilage separating the main shaft from the terminal ossified processes; also, the ossifications in the lateral processes are slender and elongate. One specimen is approximately 2.9 mm. long with a base 1.3 mm. wide.

Microtus ochrogaster.—The baculum in this species was figured and described by Hibbard and Rinker (1943) and again by Hamilton (1946). Hibbard and Rinker found only a median osseous process distal to the main shaft in a series from Kansas. Hamilton found one of seven with ossification in the lateral processes. Of three cleared and stained specimens in our collection, from South Dakota, two have ossified nodules in the lateral processes. The nodules are anterior to the distal end of the shaft, thus differing from *mexicanus*. The basal part of the shaft is broad and flattened on top, deeply notched ventrally (Pl. XXIV, m, n).

Measurements of two specimens are, respectively: length of shaft, 3.8, 3.9; width of base, 1.5, 1.8.

GENUS *Pitymys*

Pitymys pinetorum.—Two specimens from Michigan do not fit the description and figures given by Hamilton (1946) for this species. The broad angular base tapers rapidly into the shaft which is nearly parallel-sided through the middle, then expands slightly at the tip. There is a relatively wide cartilaginous connection between the shaft and the osseous terminal processes. The median process is largest; the lateral ones curve upward and outward like little horns (Pl. XXIV, k, l). This is strikingly different from that described by Hamilton (*ibid.*) where the shaft tapers from the broad base to an acute point and the osseous processes are very small. A specimen from Oklahoma and five from México (*quasiater*) do not have an expanded distal end of the shaft. They differ further from the Michigan specimens in that the base, when viewed from the end, is dumbbell-like, not convex dorsally and concave ventrally. Larger series from throughout the range of the genus might show geographic variation in this element.

Measurements of two specimens from Oklahoma (1) and Michigan (1), respectively, are: length of shaft, 2.5, 2.3; width of base, 1.4, 1.2.

GENUS *Lagurus*

Dearden (1958) recently described and illustrated bacula in this genus. They are typical of the microtines with a basal shaft and three ossified digital processes terminally. Dearden, by statistical analysis, thought he could distinguish some of the subspecies of *L. curtatus* by this element alone. I suspect that had he studied larger series some of the apparent differences might have disappeared. His mean length of shaft ranges from 1.96 (*levidensis,* 12 specimens) to 2.746 (*pallidus,* 3 specimens). His extreme measurements, same two subspecies, are 1.60 and 2.96.

GENUS *Sigmodon*

Sigmodon hispidus.—The baculum of the cottonrat is similar to those of certain microtines, particularly *Ondatra*. The broad base tapers abruptly into the laterally compressed shaft which has a bulbous distal end (Pl. XXV, g, h). I do not detect a distinct "keel" as described by Hamilton (1946) for this species. The basal portion is U-shaped when viewed from the end, the concavity is dorsal. The three digitate processes are cartilaginous until late in development, when they become ossified. The central process is rounded in cross section—it extends forward for about half its length then bends upward at a nearly right angle. The lateral ossified processes differ from any microtines studied in that they are flattened and spatulate distally. There is a cartilaginous connection between the processes and the shaft.

Measurements of 18 specimens from Florida (7), Arizona (6), Oklahoma (1), and México (4) are: length of shaft, 4.7–6.8 (5.6); width of base, 2.3–3.3 (2.9).

GENUS *Ondatra*

Ondatra zibethica.—The baculum of the muskrat is the largest seen in microtines (Pl. XXV, i, j). In general, the basal part is wide and rugose with a condyl-like structure on each side. In some specimens there is a definite median notch at the base. A depression both above and below gives it a dumbbell shape when viewed from the end. The shaft is usually curved upward, in some sharply, in others gradually. The distal end of the shaft may be a rounded point or it may be slightly expanded. The digitate processes are always ossified in adults. In most of my specimens the central one is longest and the other two are laterally compressed. Hamilton (1946) stated that the median one is shortest. The processes are connected to the shaft by cartilage.

Measurements of eight specimens from Nevada (1), Michigan (3), and North Carolina (4) are: length of shaft 5.5–6.8 (6.0); width of base, 2.5–3.8 (3.1).

FAMILY ZAPODIDAE

The two North American genera in this family, *Zapus* and *Napaeozapus*, have relatively simple bacula (Pl. I). In both, the bone curves slightly upward then downward as it gradually tapers from the base to near the tip. A median, longitudinal groove on the dorsal surface of the expanded base is more pronounced in *Zapus*, also, the tip is dorsoventrally flattened whereas it is round in *Napaeozapus*. Both were illustrated and briefly described by Krutzsch (1954).

Zapus hudsonius.—Bone as described above; averages shortest of those examined. There is little apparent variation in available material (Pl. I, a, b).

Measurements of 20 specimens from Michigan (13), South Dakota (5), Oklahoma (1), and Virginia (1) are: length, 4.3–5.1 (4.68); width of base, 0.5–0.7 (0.6); height of base, 0.3–0.5 (0.4).

Zapus princeps.—Similar to that of *hudsonius*, but slightly larger. Krutzsch (*ibid.*), in his illustration, indicates a laterally flattened tip, but in his description (p. 394) he states that the tip is "rounded and dished out in dorsal aspect . . . "

Measurements of five specimens from Montana are: length, 4.7–6.0 (5.3); width of base, 0.5–0.9 (0.7); height of base, 0.4–0.6 (0.5).

Zapus trinotatus.—I have no specimens; the following is from Krutzsch (1954), "Size large (total length 6.7 mm to 7.4 mm); base broad (0.7 mm to 0.9 mm); tip broad (0.44 mm to 0.57 mm); spade-shaped in dorsal aspect and tilted upward, gradually tapering to thin-edged tip; shaft rounded, straight." His illustration shows this species to differ from the other two principally in the much broader tip.

Napaeozapus insignis.—The bone differs from those of *Zapus* in having a round, not spatulate, distal end (Pl. I, c).

Measurements of nine specimens from Maryland (2), West Virginia (2), North Carolina (1), and Michigan (4) are: length, 5.2–6.3 (5.8); width of base, 0.5–0.9 (0.7); height of base, 0.4–0.7 (0.47).

FAMILY ERETHIZONTIDAE

Erethizon dorsatum.—There is considerable variation in detailed shape of the baculum in the porcupine. I have five specimens from Montana. They all have bulbous bases and all are flattened dorsoventrally. In two, the distal end comes to an acute point (Pl. I, e). In one, the distal end is pointed, but blunt, and in the other two the distal end is rounded. They differ in shape from all other North American rodents examined.

Measurements of the five specimens (one apparently young) are: length, 12.0–16.2 (13.7); height of base, 1.5–2.4 (2.1); width of base, 3.1–5.0 (4.3).

LITERATURE CITED

ALLEN, J. A.
1895 Descriptions of new American mammals. Bull. Amer. Mus. Nat. Hist., 7:327–40.

ARGYROPULO, A. J.
1929 Beiträge zur Kenntnis der *Murinae* Baird I, II. Zeitschr, für Säugetierkunde, 4:144–56, 15 figs.

ARNDT, RICH.
1889 Beitrag zur Anatomie und Entwickelungsgeschichte des Rutenknockens (Dissertation, Erlangen 1889) *fide* Chaine, 1925.

BENSON, SETH B.
1939 Descriptions and records of harvest mice (genus Reithrodontomys) from Mexico. Proc. Biol. Soc. Wash., 52:147–50, 1 fig.

BLAINVILLE, H. M D. DE
1839 Ostéographie ou description iconographique comparée du squelette et du système dentaire des mammifères récents cinq classes d'animaux vertébres et fossiles pour servir de base à la zoologie et à la geologie. Vol. I, Paris.

BLAIR, W. F.
1942 Systematic relationships of Peromyscus and several related genera as shown by the baculum. Jour. Mamm., 23: 196–204, 2 figs.
1954 Mammals of the mesquite plains biotic district in Texas and Oklahoma, and speciation in the central grasslands, Texas Jour. Sci., 6:235–64, 1 fig.

BOULWARE, JEAN T.
1943 Two new subspecies of kangaroo rats (Genus Dipodomys) from southern California. Univ. Calif. Publ. Zool., 46:391–96, 2 figs.

BRYANT, MONROE, D.
1945 Phylogeny of Nearctic Sciuridae. Amer. Mid. Nat., 33:257–390, 48 figs., 6 pls.

BURT, WILIAM H.
1936 A study of the baculum in the genera Perognathus and Dipodomys. Jour. Mamm., 17:145–56, 6 figs.

BURT, WILLIAM H., AND FREDERICK S. BARKALOW, JR.
1942 A comparative study of the bacula of wood rats (Subfamily Neotominae). Jour. Mamm., 23:287–97, 3 figs.

CALLERY, ROBERDEAU
1951 Development of the os genitale in the golden hamster, *Mesocricetus (Cricetus) auratus.* Jour. Mamm., 32:204–07, 1 fig.

CARUS UND OTTO
1840 Erläuterungstafeln der vergleichenden Anatomie, Leipzig, Vol. 5 (*fide* Gilbert, 1892).

CHAINE, J.
1925 L'os pénien, étude descriptive et comparative. Actes, Soc. Linn., Bordeaux, 78: 195 pp., 133 figs.

CLARK, WILLIAM K.
1953 The baculum in the taxonomy of *Peromyscus boylei* and *P. pectoralis*. Jour. Mamm., 34:189–92, 1 fig.

CUVIER, G.
1846 Leçons d'anatomie comparée, 1846, 2nd ed., Vol. 8, Paris.

DAUBENTON
1758–1767 Histoire naturelle, général et particulière avec la description du Cabinet du Roi. Paris. Vols. 7–15.

DEANSLEY, RUTH
1935 The reproductive processes of certain mammals. Part IX—Growth and reproduction in the stoat (*Mustela erminea*). Philos. Trans. Roy. Soc. London, ser. B, 225:459–92, 7 figs.,Pls. 28–31.

DEARDEN, LYLE C.
1958 The baculum in *Lagurus* and related microtines. Jour. Mamm., 39:541–53, 1 fig.

DIDIER, ROBERT
1948 Etude systématique de l'os pénien des mammifères. Mammalia, 12:67–93, 21 figs.
1950 Etude systématique de l'os pénien des mammifères. *Ibid.*, 14:78–94, 14 figs.
1952 Etude systématique de l'os pénien des mammifères. *Ibid.* 16:7–23, 14 figs.
1943 Etude systématique de l'os pénien des mammifères. *Ibid.*, 17:67–74, 7 figs.

ELDER, WILLIAM H.
1951 The baculum as an age criterion in mink. Jour. Mamm., 32:43–50, 2 pls., 3 figs.

ELLERMAN, J. R.
1940 The families and genera of living rodents. British Museum (Natural History), Vol. 1, London.

FRILEY, CHARLES E., JR.
1949a Age determination, by use of the baculum, in the river otter, *Lutra c. canadensis* Schreber. Jour. Mamm., 30:102–10, 2 pls., 1 fig.
1949b Use of the baculum in age determination of Michigan beaver. *Ibid.* 30:261–67, 1 pl.

GILBERT, TH.
1892 Das os priapi der Säugethiere. Morp. Jahrb., 18:805–31, Pl. 27.

GOLDMAN, E. A.
1912 Descriptions of twelve new species and subspecies of mammals from Panama. Smith. Miscl. Coll., 56 (36) :1–11.

HALL, E. R., AND KEITH R. KELSON
1959 The mammals of North America. New York. The Ronald Press. 2 vols.

HAMILTON, W. J., JR.
1946 A study of the baculum in some North American Microtinae. Jour. Mamm., 27:378–87, 1 pl., 3 figs.
1949 The bacula of some North American vespertilionid bats. *Ibid.*, 30:97–102, 1pl.

HERSHKOVITZ, PHILIP
1944 A systematic review of the Neotropical water rats of the genus *Nectomys* (Cricetinae). Misc. Publ. Mus. Zool., Univ. Mich., 58:101 pp., 4 pls, 2 maps, 5 figs.

HIBBARD, CLAUDE W., AND GEORGE C. RINKER
1942 A new bog-lemming (Synaptomys) from Meade County, Kansas. Univ. Kans. Sci. Bull., 28:25–35, 3 figs.
1943 A new meadow mouse (*Microtus ochrogaster taylori*) from Meade County, Kansas. *Ibid.*, 29:255–68, 5 figs.

HOFFMEISTER, DONALD F., AND LUIS DE LA TORRE
1959 The baculum in the wood rat *Neotoma stephensi*. Proc. Biol. Soc. Wash., 72: 171–72, 1 fig.

HOFFMEISTER, DONALD F., AND JOHN R. WINKELMANN
1958 The os clitoridis in the badger, Taxidea taxus. Trans. Ill. Acad. Sci. (1957), 50:233–34, 2 figs.

HOLLISTER, N.
1915 The genera and subgenera of raccoons and their allies. Proc. U. S. Natl. Mus., 49:143–50, 2 pls.

HOOPER, EMMET T.
1958 The male phallus in mice of the genus *Peromyscus*. Miscl. Publ. Mus. Zool. Univ. Mich., 105:24 pp., 14 pls., 1 fig.
1959 The glans penis in five genera of cricetid rodents. Occ. Papers, Mus. Zool. Univ. Mich., 613:1–11, 5 pls.

HOWELL, ARTHUR H.
1929 Revision of the American chipmunks (Genera Tamias and Eutamias). N. Amer. Fauna, 52:157 pp., 10 pls., 9 figs.
1938 Revision of the North American ground squirrels, with a classification of the North American Sciuridae. *Ibid.*, 56:256 pp., 32 pls., 20 figs.

JAEGER, EDMOND C.
1947 Use of the os phallus of the rac[c]oon as ripping tool. Jour. Mamm., 28:297, 1 fig.

JELLISON, W. L.
1945 A suggested homolog of the os penis or baculum of mammals. Jour. Mamm., 26:146–47.

JOHNSON, DAVID H.
1943 Systematic review of the chipmunks (Genus Eutamias) of California. Univ. Calif. Publ. Zool., 48:63–148, 1pl., 12 figs.

KENNERLY, THOMAS E., JR.
1958 The baculum in the pocket gopher. Jour. Mamm., 39:445–46, 1 fig.

KRUTZSCH, PHILIP H.
1954 North American jumping mice (Genus Zapus). Univ. Kans. Publ., Mus. Nat. Hist., 7:349–472, 47 figs.

KRUTZSCH, PHILIP H., AND TERRY A. VAUGHN
1955 Additional data on the bacula of North American bats. Jour. Mamm., 36:96–100, 1 fig.

LAWRENCE, BARBARA
  1941 Neacomys from northwestern South America. Jour. Mamm., 22:418–27, 2 figs.

LAYNE, JAMES N.
  1952 The os genitale of the red squirrel, *Tamiasciurus*. Jour. Mamm., 33:457–59, 1 fig.
  1954 The os clitoridis of some North American Sciuridae. *Ibid.*, 35:357–67, 2 figs.

LÖNNBERG, EINER
  1911 Der Penisknochen zweier seltener Carnivoren. Anat Anz. 38:230–32, 2 figs.

LOUGHREY, ALAN G.
  1959 Preliminary investigations of the Atlantic walrus, *Odobenus rosmarus rosmarus* (Linnaeus). Ottawa, Wildlife Mgt. Bull., ser. 1, no. 14: 123 pp.

MAXIMILIAN, PRINZEN ZU WIED
  1862 Verzeichniss der auf seiner Reise in Nord-Amerika beobachteten Säugethiere. Berlin, Nicolaische Verglasbuchandlung (G. Parthey), 240 pp.

MILLER, GERRIT S., JR., AND REMINGTON KELLOGG
  1955 List of North American Recent mammals. U.S. Natl. Mus. Bull., 205:xii + 954.

MILLER, MALCOLM E.
  1952 Guide to the dissection of the dog. Ithaca, New York (Lithoprinted by Edwards Brothers, Inc., Ann Arbor, Michigan):xii + 369, illustr.

MOORE, JOSEPH C.
  1959 Relationships among living squirrels of the Sciurinae. Amer. Mus. Nat. Hist., Bull., 118:153–206, 7 figs.

MOSSMAN, H. W., J. W. LAWLACH, AND J. A. BRADLEY
  1932 The male reproductive tract of the Sciuridae. Amer. Jour. Anat., 51:89–155, 16 figs., 7 pls.

MURIE, O. J.
  1936 Notes on the mammals of St. Lawrence Island, Alaska. Miscl. Publ., Univ. Alaska, 2:337–46, 3 figs.

OSGOOD, W. H.
  1909 Revision of the pocket mice of the genus Perognathus. N. Amer. Fauna, 18:72 pp., 4 pls., 15 figs.

OWEN, RICHARD
  1868 On the anatomy of vertebrates. Longmans, Green, and Co., London, Vol. 3.

PALLAS, P. S.
  1767 Spicilegia zoologica (*fide* Chaine, 1925)
  1778 Novae species quadrupedum e glirium ordine, etc. viii + 388, 38 pls.

PERRAULT, CL.
  1733 Mémoires pour servir à l'histoire naturelle des animaux. Mem. Acad. Roy. des Sci., Paris (1666–1699), vol. 3.

POCOCK, R. I.
  1918 The baculum or os penis of some genera of Mustelidae. Ann. Mag. Nat. Hist., ser. 9, vol. 1:307–12, Figs. a–m.

1921   The external characters and classification of the Procyonidae. Proc. Zool. Soc. London (1921): 389–422, 13 figs.

1923   The classification of the Sciuridae. *Ibid.* (1923):209–46, Figs. 18–29.

POHL, LOTHAR

1909   Ueber das os penis der Musteliden. Jen. Zeithschr. Naturw., 45:381–94.

1911   Das os penis der Carnivoren einschliesslich der Pinnipedier. Jen. Zeitsch. Naturw., 47:115–60, 13 figs., Pls. 7, 8.

RETTERER, M. ED.

1887   Note sur le développement du pénis et du squelette du gland chez certains rongeurs. Compt. Rend. Soc. Biol., 1887, Paris, vol. 4:496–98.

RETTERER, ED., ET H. NEUVILLE

1913   Du squelette pénien de quelques Mustélidés. Compt. Rend. Soc. Biol., 1913, Paris, vol. 75:622–24.

RINKER, GEORGE C.

1944   Os clitoridis from the rac[c]oon. Jour. Mamm., 25:91–92, 1 fig.

1954   The comparative myology of the mammalian genera *Sigmodon, Oryzomys, Neotoma,* and *Peromyscus* (Cricetinae), with remarks on their intergeneric relationships. Misc. Publ., Mus. Zool., Univ. Mich., 83:124 pp., 18 figs.

RUTH, E. B.

1934   The os priapi: a study in bone development Anat. Record, 60:231–44, 3 pls.

SCHEFFFER, VICTOR B.

1939   The os clitoridis of the Pacific otter. Murrelet, 20:20–21, 6 figs.

1942   An on clitoridis from aplodontia. Jour. Mamm., 23:443, 1 fig.

1949   The clitoris bone in two pinnipeds. *Ibid.,* 30:269–70, 1 pl.

1950   Growth of the testes and baculum in the fur seal, *Callorhinus ursinus. Ibid.,* 31:384–94, 3 pls., 2 figs.

SCHWARTZ, ALBERT, AND EUGENE P. ODUM

1957   The woodrats of the eastern United States. Jour. Mamm., 38:197–206, 2 figs.

SETZER, HENRY W.

1949   Subspeciation in the kangaroo rat, Dipodomys ordii. Univ. Kans. Publ., Mus. Nat. Hist., 1:473–573, 27 figs.

SIERTS, WERNER

1950   Os clitoridis von Zalophus californianus Less. und Sciurus vulg. fuscoater Altum. Zool. Anz., Leipzig, 145:938–39.

SIMOKAWA, SEIMATU

1938   Einige Bermerkungen über den Clitorisknochen. Keijo Jour. Med., 9:273–82, 2 figs.

SIMPSON, G. G.

1945   The principles of classification and a classification of mammals. Amer. Mus. Nat. Hist., Bull., 85:xvi + 350.

SPRAGUE, JAMES M.

1939   A preliminary study of the baculum of *Peromyscus leucopus* and *P. maniculatus* in Kansas. Trans. Kan. Acad. Sci., Topeka, 42:495–97, Figs. 2–3.

TAMSITT, J. R.
1958 The baculum of the *Peromyscus truei* species group. Jour. Mamm., 39:598–99, 1 fig.

THOMAS, O.
1915 The penis-bone, or 'baculum," as a guide to the classification of certain squirrels. Ann. Mag. Nat. Hist., London, 15:383–87.

TOBINGA, SUEKI
1938 Beitrag zur Kenntnis der Morphologie des Penisknochens. Keijo Jour. Med. 9:244–72, 1 pl.

TULLBERG, TYCHO
1899 Ueber das System der Nagethiere, eine phylogenetische Studie. Upsala. PP. i–vi, 1–514, A 1–A18, 57 pls.

VINOGRADOV, B. S.
1925 On the structure of the external genitalia in Dipodidae and Zapodidae (Rodentia) as a classificatory character. Proc. Zool. Soc., London, (1925):577–85.

WADE, O., AND P. T. GILBERT
1940 The baculum of some Sciuridae and its significance in determining relationships. Jour. Mamm., 21: 52–63, 3 figs.

WHITE, JOHN A.
1953a Genera and subgenera of chipmunks. Univ. Kan. Publ., Mus. Nat. Hist., 5:543–61, 12 figs.
1953b The baculum in the chipmunks of western North America. *Ibid.* 611–31, 19 figs.

WRIGHT, PHILIP L.
1947 The sexual cycle of the male long-tailed weasel (*Mustela frenata*). Jour Mamm., 28:343–52, 1 pl., 4 figs.

Accepted for publication January 4, 1960

PLATES I–XXV

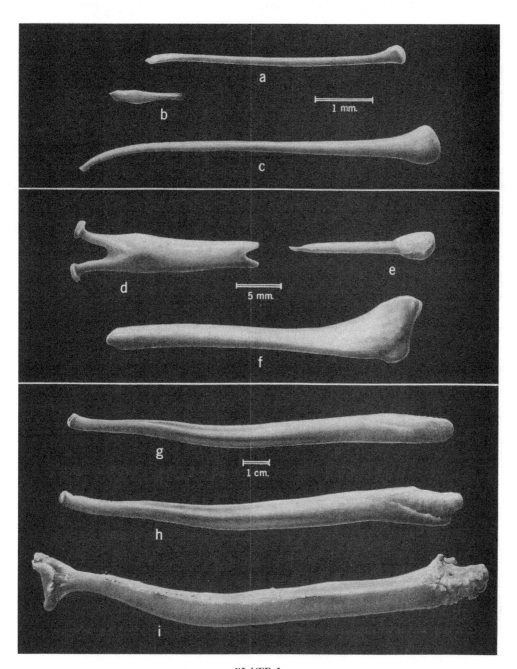

PLATE I

Miscellaneous bacula: a, b, *Zapus hudsonius* (P. 3334); c, *Napaeozapus insignis* (P. 1954); d, *Aplodontia rufa* (P. 1977); e, *Erethizon dorsatum* (P. 2338); f, *Castor canadensis* (P. 3280); g, *Ursus americanus* (47281); h, *U. sitkensis* (97532); i, *Zalophus californianus* (P. 2354).

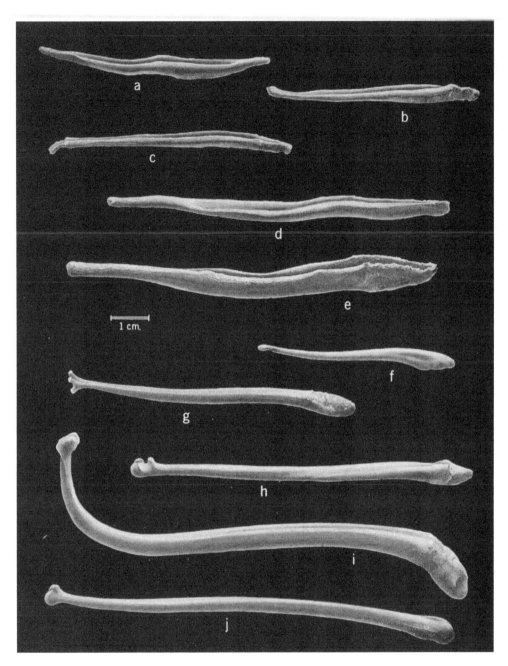

PLATE II

Bacula of carnivores: a, *Urocyon cinereoargenteus* (71879); b, *Vulpes fulva* (87125); c, *Alopex lagopus* (P. 3759); d, *Canis latrans* (82493); e, *C. lupus* (60942); f, *Bassariscus astutus* (P. 3812); g, *Potos flavus* (79521); h, *Nasua narica* (63160); i, *Procyon lotor* (61686); j, *P. cancrivorus* (46409).

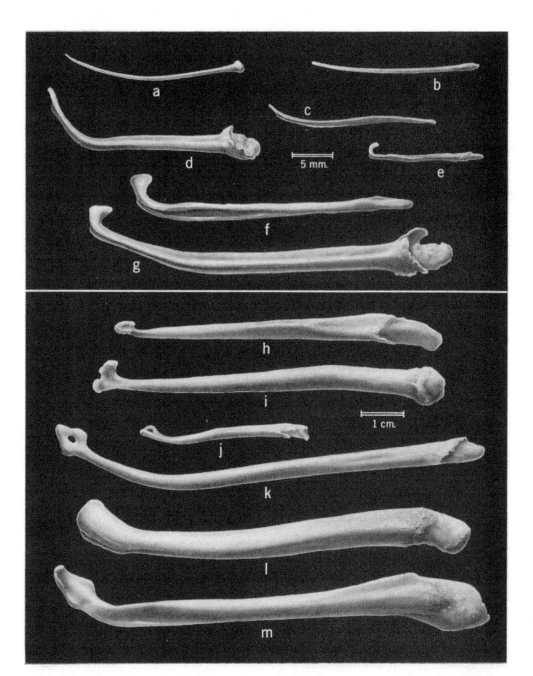

PLATE III

Bacula of mustelids: a, *Spilogale putorius* (86492); b, *Mephitis mephitis* (78553); c, *Mustela erminea* (83611); d, *M. frenata* (P. 2816); e, *M. rixosa* (88078); f, *M. nigripes* (103451, juv.); g, *M. vison* (98008); h, *Eira barbara* (64937); i, *Gulo luscus* (98096); j, *Martes americana* (97958); k, *M pennanti* (97981); l, *Lutra canadensis* (86273); m, *Taxidea taxus* (54724).

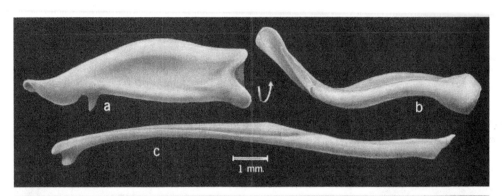

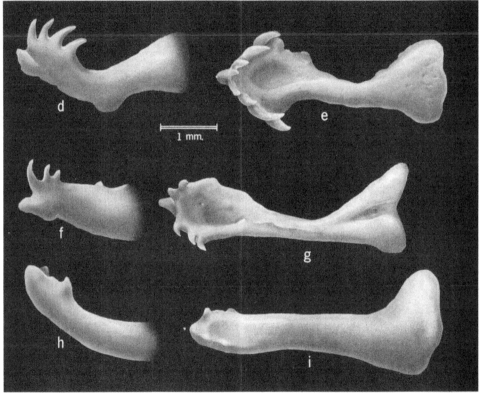

PLATE IV

Bacula of sciurids: a, b, *Glaucomys sabrinus* (84434); c, *G. volans* (P. 183); d, e,
Cynomys gunnisoni (81973); f, g, *C. ludovicianus* (P. 2318); h, i, *Marmota flaviventris* (P.
2332); j, k. *Citellus lateralis* (P. 2323).

PLATE V

Bacula of ground squirrels: a, b, *Citellus variegatus* (81980); c, d, *Ammospermophilus leucurus* (P. 131); e, f, *C. annulatus* (80981); g, h, *C. tereticaudus* (87896); i, j, *C. frank-lini* (53288); k, l, *C. columbianus* (P. 2302).

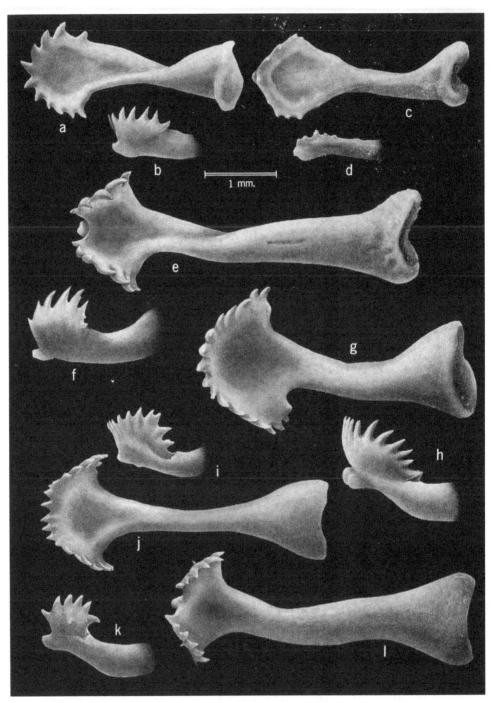

PLATE VI

Bacula of ground squirrels: a, b, *C. elegans* (P. 2313); c, d, *C. richardsoni* (87351); e, f, *C. mexicanus* (79346); g, h, *C. spilosoma* (85409); i, j, k, l, *C. tridecemlineatus* (i, j, 93214 S. Dakota; k, l, P. 3207 Mich.).

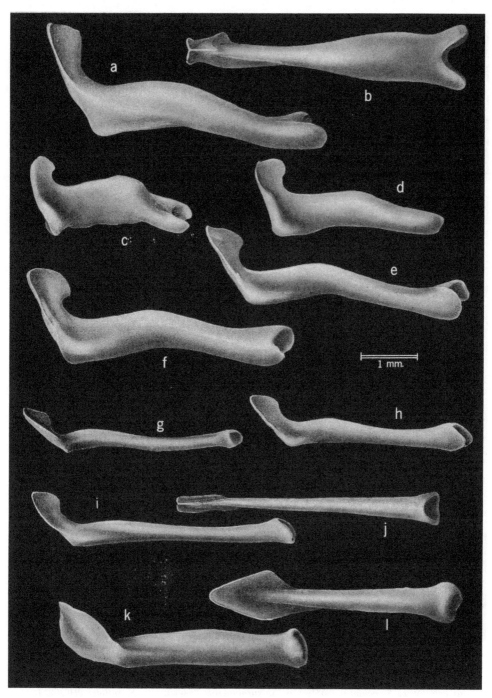

PLATE VII

Bacula of chipmunks: a, b, *Eutamis bulleri* (99973); c, *E. speciosus* (P. 150); d, *E. umbrinus* (62657); e, *E. quadrivittatus* (82030); f, *E. ruficaudus* (53706); g, *E. minimus* (62660); h, *E. amoenus* (62652); i, j, *E. dorsalis* (82009); k, l. *Tamias striatus* (82594).

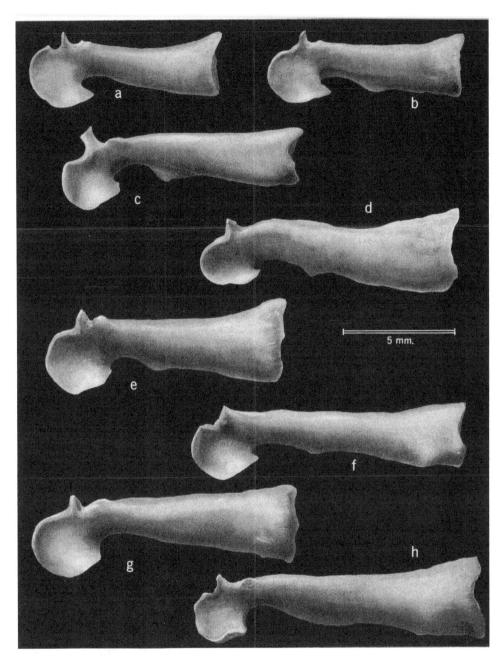

PLATE VIII

Bacula of tree squirrels: a, *Sciurus (Microsciurus) alfari* (62844); b ,*Sciurus granatensis* (62846); c, *S. oculatus* (96315); d, *S. yucatanensis* (95680); e, *S. colliaei* (94636); f, *S. arizonensis* (78011); g, *S. poliopus* (91896); h, *S. variegatoides* (62694).

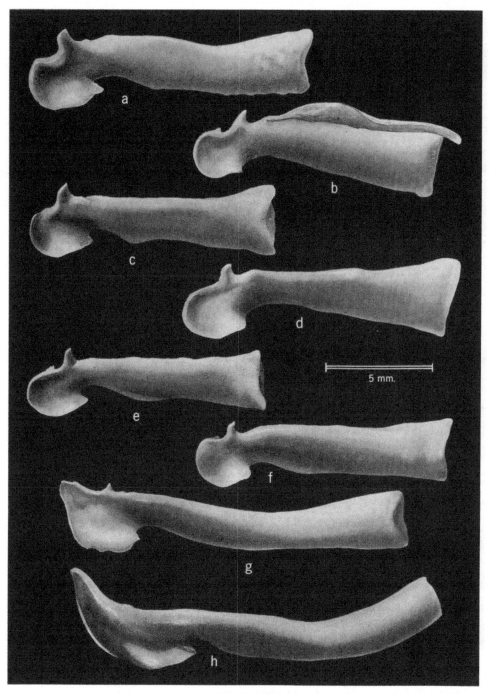

PLATE IX

Bacula of tree squirrels: a, *S. niger* (91041); b, *S. nelsoni* (88645); c, *S. alleni* (88643); d,
S. socialis (80996); e, *S. negligens* (93599); f, *S. carolinensis* (84431); g, *S. aberti* (81986);
h, *S. griseus* (P. 211).

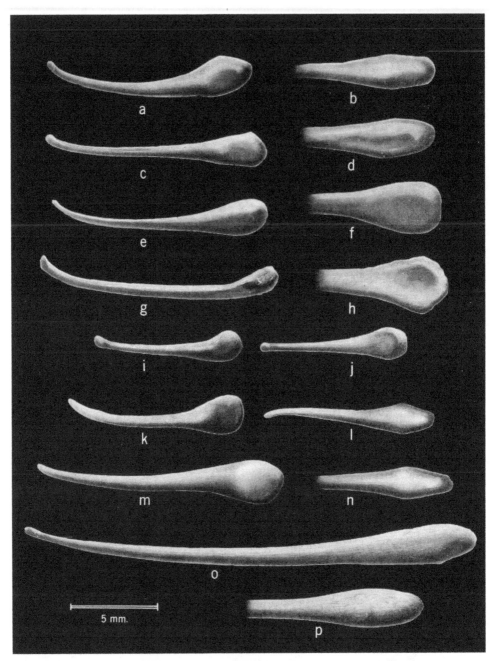

PLATE X

Bacula of pocket gophers: a, b, *Geomys bursarius* (78017); c, d, *Zygogeomys trichops* (89488); e, f, *Cratogeomys varius* (89487); g, h, *C. merriami* (91710), i, j, *Thomomys bulbivorus* (105652); k, l, *T. umbrinus* (DRD, 18905); m, n, *T. bottae* (82145); o, p, *T. talpoides* (82137).

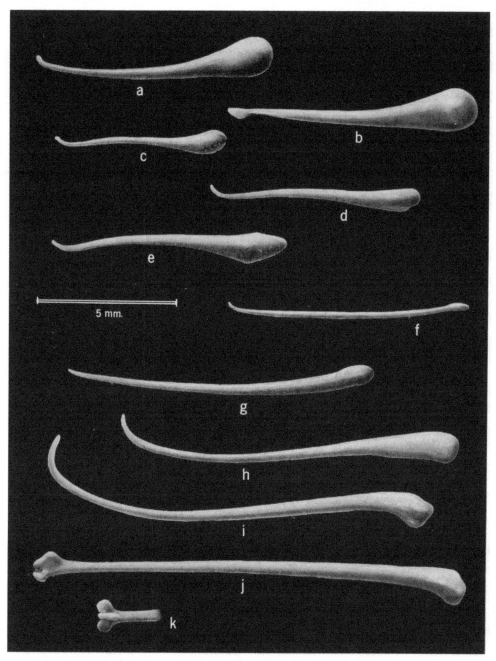

PLATE XI

Bacula of pocket mice: a, *Liomys irroratus* (96616); b, *L. pictus* (80714); c, *Perognathus merriami* (78113); d, *P. fasciatus* (87507); e, *P. parvus* (86549); f, *P. formosus* (P. 695); g, *P. baileyi* (P. 697); h, *P. pennicillatus* (61662); i, *P. goldmani* (DRD, 51089); j, k, *P. hispidus* (78108).

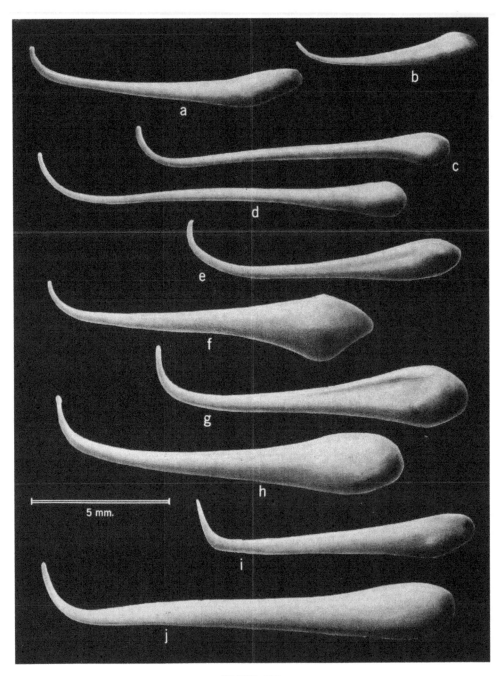

PLATE XII

Bacula of kangaroo mice and rats: a, *Dipodomys deserti* (87954); b, *Microdipodops pallidus* (78772); c, *Dipodomys merriami* (84933); d. *D. nitratoides* (DRD, 50674); e, *D. agilis* (92964)); f, *D. heermanni* (DRD, 10029); g, *D. ordi* (76023); h, *D. ingens* (DRD, 10051); i, *D. phillipsi* (88652); j, *D. spectabilis* (87034).

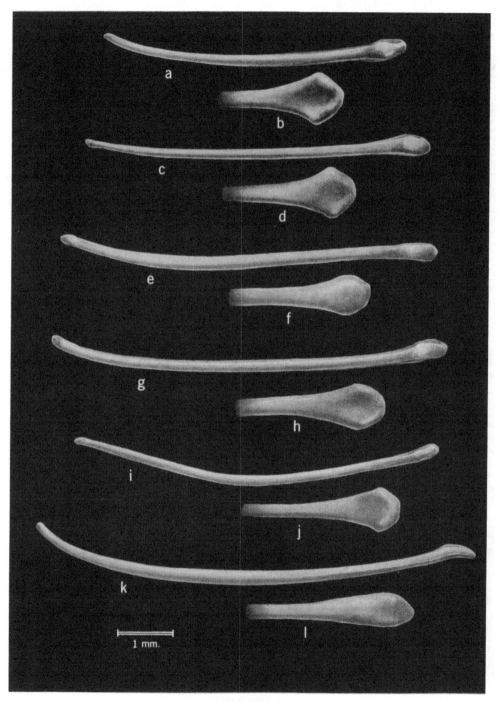

PLATE XIII

Bacula of harvest mice: a, b, *Reithrodontomys montanus* (76087); c, d, *R. humulis* (88149); e, f, *R. creper* (AMNH, 141201); g, h, *R. chrysopsis* (91812); i, j, *R. mexicanus* (DRD 10903); k, l, *R. sumichrasti* (91487).

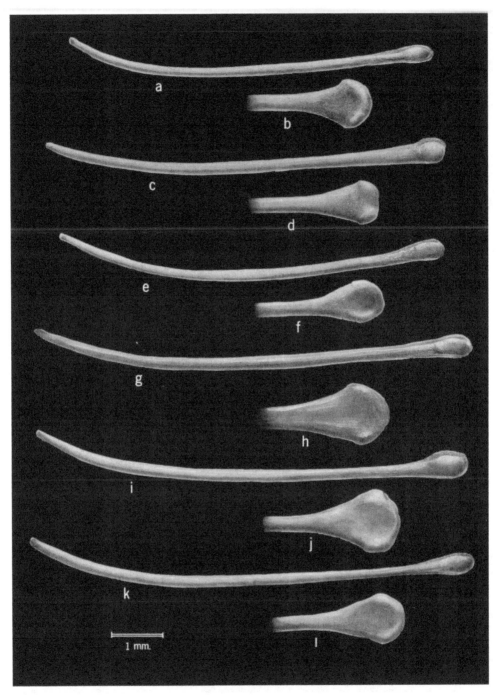

PLATE XIV

Bacula of fulvous harvest mice, *Reithrodontomys fulvescens*: a, b, Michoacán (91835);
c, d, Guerrero (90712); e, f, Hidalgo (91707); g, h, Arizona (DRD, 18765); i, j. Puebla
(91457); k, l, Michoacán (91834).

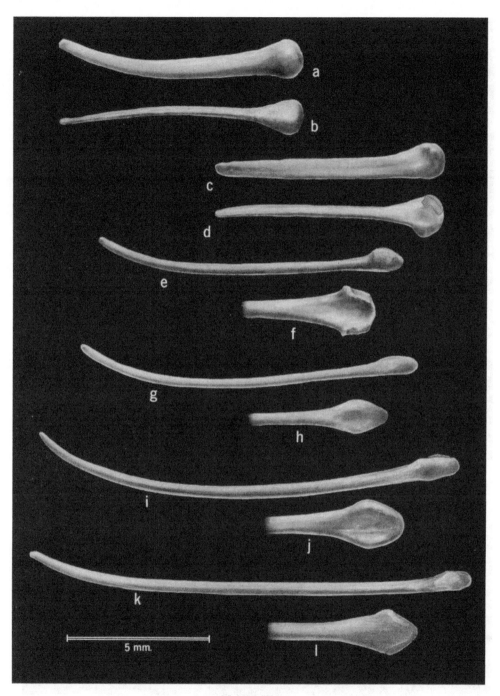

PLATE XV

Bacula of grasshopper mice and harvest mice: a, b, *Onychomys leucogaster* (76042); c, d, *O. torridus* (79258); e, f, *Reithrodontomys gracilis* (65220); g, h, *R. megalotis* (86587, Washington); i, j, *R. megalotis* (88822, Mexico); k, l, *R. microdon* (91815).

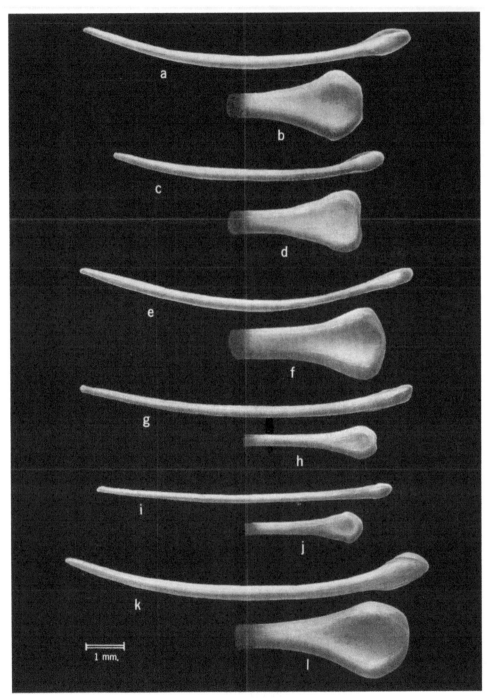

PLATE XVI

Bacula of *Peromyscus*: a, b, *P. crinitus* (88035); c, d, *P. collatus* (DRD, 50438); e, f,
P. pseudocrinitus (DRD, 50333); g, h, *P. caniceps* (DRD, 50274); i, j, *P. eremicus eva*
(80867); k, l, *P. e. eremicus* (78285).

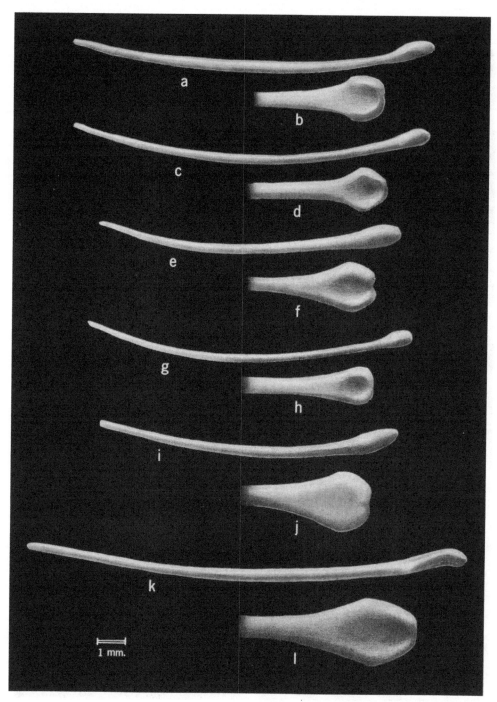

PLATE XVII

Bacula of *Peromyscus*: a, b, *P. sleveni* (P. 1444); c, d, *P. stephani* (P. 1004); e, f, *P. g. guardia* (P. 1073); g, h, *P. g. mejiae* (P. 1077); i, j, *P. pembertoni* (P. 1009); k, l, *P. californicus* (P. 1082).

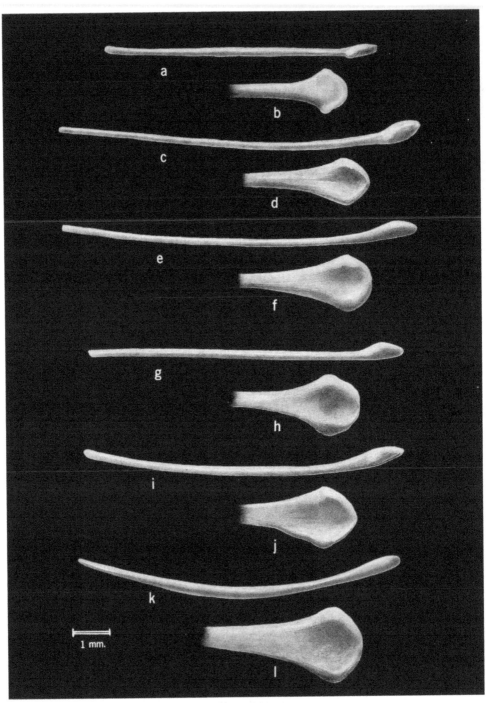

PLATE XVIII

Bacula of *Peromyscus maniculatus:* a, b, *P. m. bairdi* (82698); c, d, *P. m. oreas* (86656); e, f, *P. m. gracilis* (82683); g, h, *P. m. labecula* (90713); i, j, *P. m. exiguus* (80840); k, l, *P. m. margaritae* (80854)

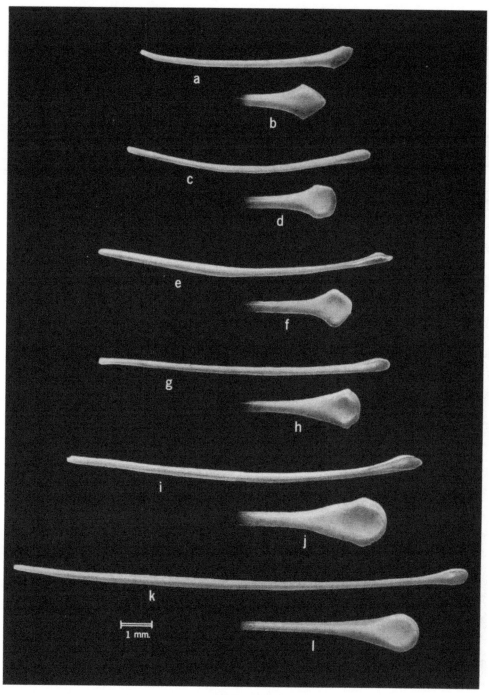

PLATE XIX

Bacula of *Peromyscus*: a, b, *P. mekisturus* (88967); c, d, *P. melanotis* (89020); e, f, *P. sejugis* (P. 1340); g, h, *P. leucopus* (84650); i, j, *P. gossypinus* (83009); k, l, *P. boylei* (88972).

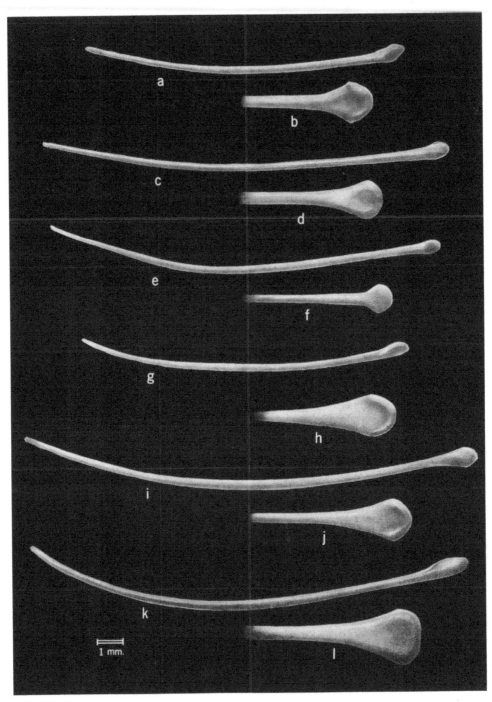

1 mm.

PLATE XX

Bacula of *Peromyscus*: a, b, *P. truei* (P. 1359); c, d, *P. nasutus* (P. 1414); e, f, *P. mexicanus* (P. 2775); g, h, *P. pectoralis* (96422); i, j, *P. melanophrys* (P. 2769); k. l. *P. difficilis* (P. 2763).

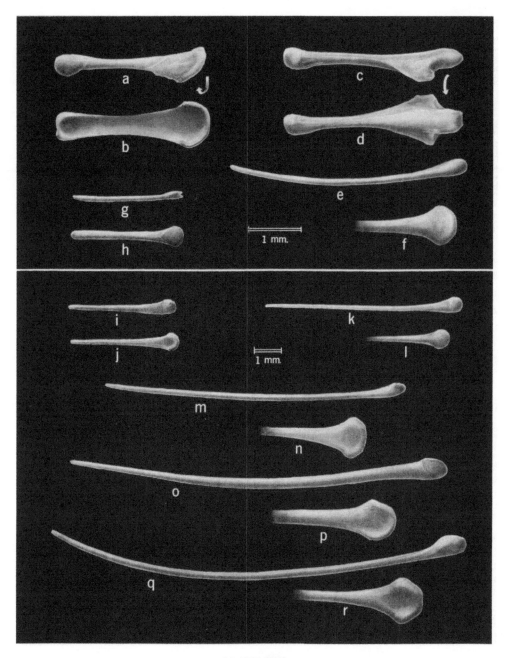

PLATE XXI

Bacula of cricetids: a, b, *Baiomys musculus* (94963); c, d, *B. musculus* (96687); e, f, *Neotomodon alstoni* (95423); g, h, *Peromyscus (Ochrotomys) nuttalli* (93497); i, j, *P. banderanus* (95384); k, l, *P. floridanus* (100868); m, n, *P. hylocetes* (95289); o, p, *P. perfulvus* (92153); q, r, *P. yucatanicus* (93335).

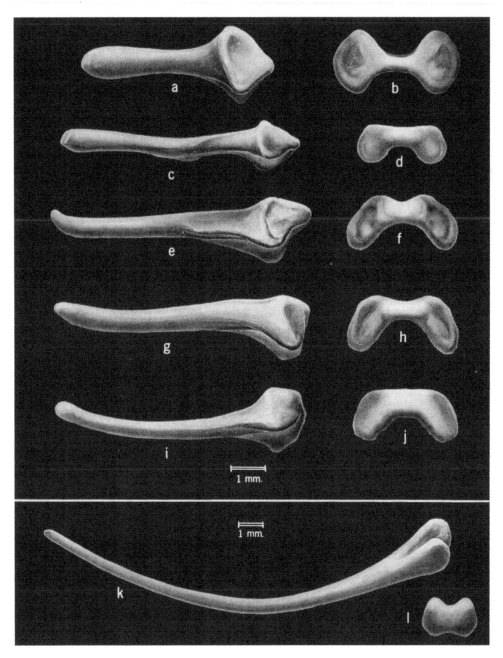

PLATE XXII

Bacula of woodrats: a, b, *Neotoma fuscipes* (P. 1784); c, d, *N. mexicana* (P. 1777); e, f, *N. floridana* (P. 2819); g, h, *N. micropus* (P. 1726); i, j, *N. albigula* (P. 1651); k, l, *N. lepida* (P. 1735).

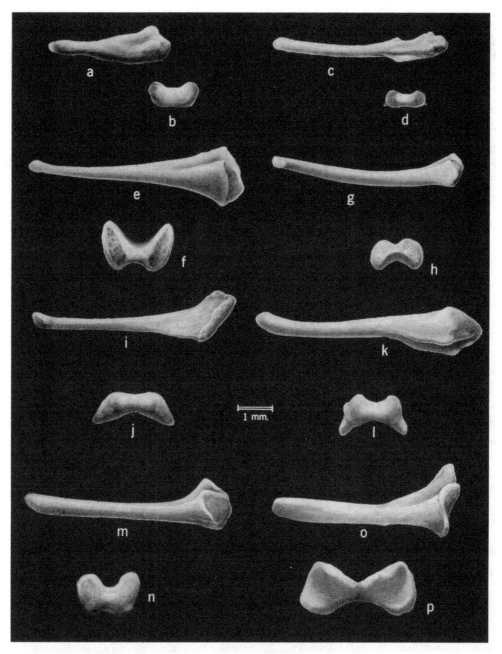

PLATE XXIII

Bacula of woodrats and spiny rats; a, b, *Neotoma phenax* (P. 1683); c ,d, *Neotoma ferruginea* (95519); e, f, *Ototylomys phyllotis* (93350); g, h, *Nyctomys sumichrasti* (P. 1817); i, j, *Neotoma alleni* (P. 1682); k, l, *N. torquata* (P. 2807); m, n, *N. cinerea* (P. 1663); o, p, *Nectomys squamipes* (CMNH 18548).

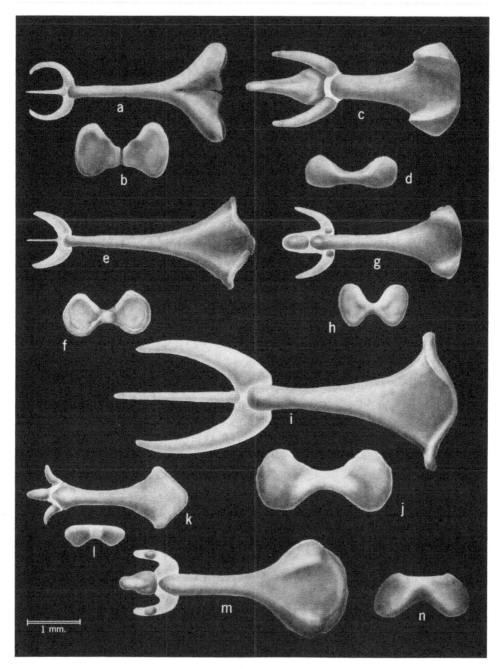

PLATE XXIV

Bacula of rice rats and voles: a, b, *Oryzomys alfaroi* (103178); c, d, *Microtus penn-sylvanicus* (P. 1870); e, f, *Oryzomys melanotis* (P. 1837); g, h, *Microtus mexicanus* (103233); i, j, *Oryzomys couesi* (94030); k, l, *Pitymys pinetorum* (P. 2195); m, n, *Microtus ochrogaster* (93270).

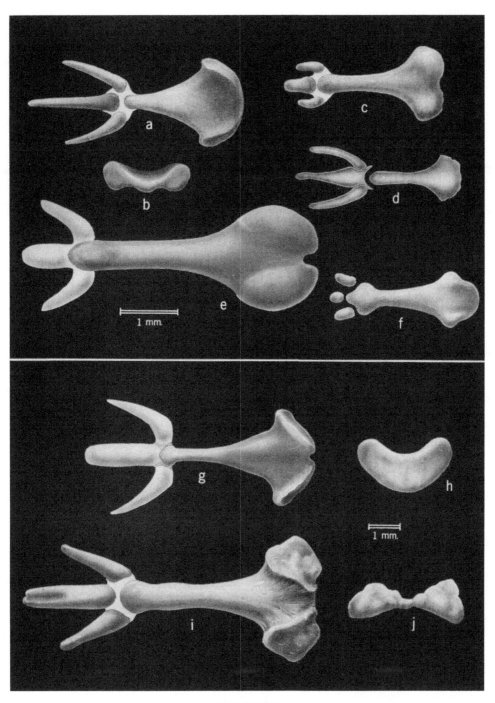

PLATE XXV

Bacula of voles, cotton rat, and muskrat: a, b, *Clethrionomys gapperi* (100948); c, *Synaptomys cooperi* (87062); d. *Lemmus helvolus* (after Hamilton, 1946): e, *Dicrostonyx hudsonius* (100767); f, *Phenacomys longicaudus* (102475); g, h, *Sigmodon hispidus* (P. 1792); i, j, *Ondatra zibethica* (P. 1817).

Printed and bound by CPI Group (UK) Ltd, Croydon, CR0 4YY

13/04/2025

14656539-0001